# SpringerBriefs in Electrical and Computer Engineering

## SpringerBriefs in Speech Technology

**Subseries Editors**
Amy Neustein

For further volumes:
http://www.springer.com/series/10043

K. Sreenivasa Rao · Sourjya Sarkar

# Robust Speaker Recognition in Noisy Environments

 Springer

K. Sreenivasa Rao
School of Information Technology
Indian Institute of Technology
Kharagpur, West Bengal
India

Sourjya Sarkar
Indian Institute of Technology Kharagpur
Kharagpur, India

ISSN 2191-737X          ISSN 2191-7388 (electronic)
ISBN 978-3-319-07129-9  ISBN 978-3-319-07130-5 (eBook)
DOI 10.1007/978-3-319-07130-5
Springer Cham Heidelberg New York Dordrecht London

Library of Congress Control Number: 2014940386

Printed on acid-free paper

Springer is part of Springer Science+Business Media (www.springer.com)

# Preface

Speaker verification (SV) is the process of validating the claimed identity of an individual using his/her speech. State-of-the-art SV systems perform reasonably well when the spoken utterances are 'clean', i.e., free from any sort of distortions caused by external factors. However, the accuracy of such systems degrade severely when speech signals are distorted due to the presence of environmental or background noise. Robustness towards environmental noise is crucial for several SV applications, especially in hand-held devices where the background environments are uncontrolled, time-varying and unpredictable. The strategies used for handling background noise can be broadly categorized as (i) 'compensation' or 'adaptation' methods where features extracted in the test environment are denoised (compensated) to match with the training environment, or speaker model parameters estimated in the training environment are altered (adapted) to reflect the test environment, and (ii) developing speaker models or extracting features which are relatively immune towards the effect of background noise, by design.

This book explores novel methods from each of the aforementioned categories for robust SV in noisy background environments. A group of Gaussian mixture model (GMM) based stochastic feature compensation methods is proposed for SV in noisy environments. Alternatively, the robustness of GMM supervector-based speaker modeling approaches is explored for SV in noisy environments. A discriminative framework is used in which fixed-size vectors obtained by stacking GMM means (i.e., supervectors) or by total variability modeling (i.e., i-vectors) are in turn used for training speaker-specific support vector machines (SVMs). These SVMs are evaluated in noisy test environments. Training an ensemble of robust SVM classifiers using adaptive boosting is proposed for improving SV performance in noisy environments.

This book is mainly intended for researchers working on robust speaker recognition technologies. This book is also useful for the young researchers, who want to pursue research in speech processing with an emphasis on acoustic modeling and feature compensation. Hence, this may be recommended as the text or reference book for the postgraduate level advanced speech processing course. The book has been organized as follows:

Chapter 1 introduces the concept of speaker recognition (SR) and the various stages involved in the SR process. The application of robustness required for SR technologies has been emphasized. Chapter 2 provides a review of the diverse array of methods employed for robust speaker/speech recognition. Chapter 3 discusses the baseline SV systems developed using Gaussian mixture models. Chapter 4 explores stochastic feature compensation methods for robust SV in noisy environments. Chapter 5 explores robust speaker modeling methods for SV in noisy environments. Chapter 6 provides a brief summary and conclusion of the book with directions towards the scope for possible future work.

We would especially like to thank all professors of the School of Information and Technology, IIT Kharagpur, for their moral encouragement and technical discussions during the course of editing and organization of this book. Special thanks to our colleagues at Indian Institute of Technology, Kharagpur, India, for their cooperation to carry out the work. We are grateful to our parents and family members for their constant support and encouragement. Finally, we thank all our friends and well-wishers.

Kharagpur, India                                                             K. Sreenivasa Rao
                                                                             Sourjya Sarkar

# Contents

# Acronyms

AANN      Auto-associative neural network
ASR      Automatic speaker recognition
AdaBoost      Adaptive boosting
BM      Background model
CDCN      Codeword dependent cepstral normalization
CMS      Cepstral mean subtraction
CMLLR      Constrained maximum likelihood linear regression
CVN      Cepstral variance normalization
FCDCN      Fast codeword dependent cepstral normalization
DCT      Discrete cosine transform
DCF      Detection cost function
DFT      Discrete Fourier transform
EER      Equal error rate
EM      Expectation maximization
FFNN      Feed-forward neural network
GMM      Gaussian mixture models
GUMI      GMM-UBM mean interval
HMM      Hidden Markov model
Hz      Hertz
IDFT      Inverse discrete Fourier transform
JFA      Joint factor analysis
KL div      Kullback Liebler divergence
LDA      Linear discriminant analysis
LSF      Line spectral frequencies
LPCC      Linear prediction cepstral coefficients
MEMLIN      Multi-environment model based linear normalization
MFCC      Mel frequency cepstral coefficients
MinDCF      Minimum detection cost function
MAP      Maximum a posteriori
MHEC      Mean Hilbert envelope coefficient
MLE      Maximum likelihood estimate

| | |
|---|---|
| MLLR | Maximum likelihood linear regression |
| MMCN | Multivariate model based cepstral normalization |
| MMSE | Minimum mean squared error |
| ms | Milliseconds |
| NAP | Nuisance attribute projection |
| NAT | Noise adaptive training |
| NIST-SRE | National Institute of Standards and Technology-Speaker Recognition Evaluation |
| NMCC | Normalized modulation cepstral coefficients |
| PLP | Perceptual linear prediction |
| PMC | Parallel model combination |
| PNCC | Power normalized cepstral coefficients |
| PPCA | Probabilistic principal component analysis |
| RASTA | Relative spectra |
| RATZ | Multivariate Gaussian based cepstral normalization |
| SE | Spectral equalization |
| SFC | Stochastic feature compensation |
| SI | Speaker identification |
| SMS | Speaker model synthesis |
| SNR | Signal-to-noise ratio |
| SR | Speaker recognition |
| SPLICE | Stereo piecewise-linear compensation for environment |
| SS | Spectral subtraction |
| SSM | Stereo-based stochastic mapping |
| SVM | Support vector machines |
| SV | Speaker verification |
| TRAJMAP | Trajectory based stochastic mapping |
| UBM | Universal background model |
| UP-AVR | Utterance partitioning with acoustic vector resampling |
| VAD | Voiced activity detection |
| VOP | Vowel onset point |
| VQ | Vector quantization |
| VTL | Vocal tract length |
| VTS | Vector Taylor series |
| WCCN | Within class covariance normalization |

# Chapter 1
# Introduction

**Abstract** This chapter introduces the concept of speaker recognition (SR) and its applications. It emphasizes on explaining the requirement of developing SR technologies that are robust towards background environments. The intermediate sections provide broad overviews of various stages associated in developing a SR system and different categories of SR. The later sections highlight the issues addressed in the book and its contributions.

## 1.1 Introduction

Telecommunication networking has made a pervasive impact in the human society in the last few decades. Much of our personal information today, is shared over the Internet or exchanged through hand-held devices. This obviously drives the demand for technology that secures human access to confidential data. Recent developments in the area of remote transactions such as telebanking, e-commerce, online railway or airline reservations etc., have made individual authentication a crucial factor. Traditional modes of security such as passwords and personal identification numbers (credit/debit cards) are often vulnerable since they can be easily forgotten, misplaced or stolen. A feasible alternative is the use of biometric authentication i.e., identifying individuals by their physical traits, which are least susceptible to physical misuse and impersonation. However, practical use of common biometric techniques like iris, face and fingerprint recognition is constrained by factors like close proximity/direct contact with individuals or requirement of costly sensors, which thereby limits their application in remote operations.

Speaker recognition (SR) is the task of recognizing individuals using their speech. As the most common mode of human communication, speech is readily available, can be easily recorded by inexpensive devices and transmitted over long-distance telecommunication channels. This is evident from the wide range of voice communication applications available over the Internet e.g., Skype, Google talk, Google voice search etc. As such, speaker recognition also provides an

K.S. Rao and S. Sarkar, *Robust Speaker Recognition in Noisy Environments*,
SpringerBriefs in Electrical and Computer Engineering,
DOI 10.1007/978-3-319-07130-5_1, © The Author(s) 2014

attractive biometric alternative to its sophisticated counterparts. Speaker recognition technologies are being readily deployed today in three major areas of applications i.e., security, surveillance and forensics [1].

The key applications that demand biometric security based SR technology are tele-commerce and forensics [1] where the objective is to automatically authenticate speakers of interest using his/her conversation over a voice channel (telephone or wireless phone). In forensics (e.g., criminal investigation), the speakers can be considered non-cooperative as they do not specifically wish to be recognized. On the other hand, in telephone-based services and access control, the users are considered to be cooperative. With the ever increasing popularity in multimedia web-portals (e.g., Facebook and Youtube), large repositories of archived spoken documents such as TV broadcasts, teleconference meetings, and personal video clips can be accessed through the Internet. Searching for topic of discussion, participant names and genders from these multimedia documents would require automated technology like speaker verification and recognition.

While the SR technologies promise an additional biometric layer of security to protect the user, the practical implementation of such systems faces many challenges. For example, a handheld-device based recognition system needs to be robust to noisy environments, such as office, street or car environments, which are subject to unpredictable and unknown sources of noise (e.g., abrupt interference, sudden environmental change, etc.).

## 1.2  Speaker Recognition

Human beings can reliably recognize known voices by barely hearing a few seconds of speech. The uniqueness of one's voice can be attributed to both physical and acquired characteristics of a person. Physical differences occur largely due to the distinct shapes and sizes of the voice producing organs (e.g., vocal folds, vocal tract, larynx, etc.) and partly due to the articulators (e.g., tongue, teeth, lip etc.). Apart from these anatomical properties, individuals can also be distinguished by their accent, vocabulary, speaking rate and other personal mannerisms that are acquired over a period of time. State-of-the-art speaker recognition systems exploit these properties in parallel to achieve high recognition accuracy [2, 3]. While subjective tests have revealed that humans often show superior performance in recognizing familiar [4] or disguised voices [5], machines outperform humans when it comes to recognition on a large scale [6] especially for non-cooperative speakers. Automatic speaker recognition (ASR) systems would ideally imitate the human voice recognition process which in turn is dependent on a complex auditory perception mechanism. Human beings are inherently capable of integrating a wide range of knowledge sources in speech signals at various levels (e.g., acoustic, articulatory, syntactic etc.). However, the exact nature of speech comprehension or segregation of speaker information at the cognitive or neurobiological level is still largely unknown. Thus, the general approach is to enumerate perceptual cues used by humans at various levels and estimate their patterns for later classification. The broad stages of the ASR process are briefly outlined in the following paragraphs.

- **Preprocessing:** This stage corresponds to the acquisition of a speech signal for the recognition process. The analog speech signal is digitized by sampling it at a desired frequency. The digital speech is usually 'pre-emphasized' using a high pass filter which emphasizes higher frequency components and compensates for the human speech production mechanism which tends to attenuate them. For several ASR tasks, a 'voiced activity detection' (VAD) stage is often used to separate speech segments from a given audio signal. It is often challenging to implement VAD that works consistently across various background environments especially for short-duration utterances [2].
- **Feature Extraction:** This stage corresponds to the enumeration of knowledge sources in a speech signal. The raw speech signal is reduced to a set of parameters in which speaker-discriminative properties are emphasized and redundant information is suppressed. The vast numbers of features explored for ASR tasks can be broadly categorized as spectral, source, prosodic and high-level features. The first two categories, often collectively termed as 'low-level' features, convey physiological information about the speaker (e.g., size of vocal folds, structure of vocal tract etc.). The latter two categories comprise high-level features which reflect acquired behavioral aspects of a speaker (e.g., temperament, accent, vocabulary etc.). Selection of appropriate features for ASR is usually based on certain criterion [7]. An ideal feature is expected to have high inter-speaker variability, low intra-speaker variability, natural occurrence in speech, robustness towards noise/channel-distortion, immunity towards a speaker's health/mood fluctuations and ease of extraction. Apart from these, the features should have a compact representation to avoid requirement of a large amount of training data. Though short-term spectral features (e.g., MFCC) [8] are often preferred for ASR tasks due to their high accuracy and real-time extraction, they are susceptible to noise degradation [9]. High-level features improve noise/channel-robustness at the cost of a difficult extraction procedure and high amount of training data. Feature selection is thus a tradeoff between speaker-discrimination, robustness and practical application.
- **Acoustic Speaker Modeling:** In this stage various statistical modeling techniques are employed to capture the distribution of features extracted from individual speakers. The feature extraction and speaker modeling stage jointly represent the training or enrollment phase of ASR in which speakers register/enrol for the SR system. The goal of this stage is to build unique templates or models for each enrolled speaker. Standard speaker modeling techniques can be categorized in different ways. Depending on the nature of modeling the feature distribution, they may be either *parametric* or *non-parametric*. *Parametric* models assume a fixed probability density of the feature distribution (e.g., Gaussian Mixture Models (GMMs) [10, 11], Hidden Markov Models (HMMs) [12]) whereas *non-parametric* models use non-stochastic template-based modeling (e.g., Vector Quantization (VQ) [13], Dynamic Time Warping [14]). On the

basis of their training paradigm, speaker models are classified as *generative* and *discriminative*. The *generative* models individually estimate feature distribution within each speaker (class) (e.g., GMMs, HMMs, VQ) while *discriminative* models are based on learning the differences between enrolled speakers (classes) (e.g., Support Vector Machines (SVMs) [15], Neural Networks (NNs) [16]). Recent research trends have also focussed on combining generative and discriminative models for improved ASR tasks [15, 17, 18].

- **Pattern Matching and Classification:** In this stage an unknown (test) utterance based on its statistical similarities with a known speaker model. The pattern matching and classification stage is collectively termed as the testing/evaluation phase in which the ASR system is evaluated on the basis of its classification accuracy. Pattern matching is entirely dependent on the nature of the acoustic speaker models. In case of stochastic generative models, matches are quantified in the form of log-likelihood scores whereas for parametric ones they might be simple distance metrics (e.g., Euclidean distance for VQ). For discriminative models, scores may be based on the distance from the decision boundary of two classes (speakers) (e.g., SVMs) or the difference between the actual and predicted class (e.g., NNs). A decision is taken based on the scores obtained i.e., the test utterance is classified as the speaker (model) producing the highest score.

## 1.3 Types of Speaker Recognition

Speaker Recognition can be broadly categorized into two types i.e., Speaker Identification (SI) [10] and Speaker Verification (SV) [11].

### 1.3.1 Speaker Identification

Closed-set speaker identification (SI) is the task of detecting a unique speaker responsible for producing a test utterance, out of a closed-set of enrolled speakers. In case the test utterance doesn't belong to any member of the closed-set, the task is an 'Open-set' SI. Considering each speaker model as a class, the SI task is basically a multi-class classification problem in which an unknown test utterance is assigned to a particular class. Figure 1.1 shows the block-diagram of a SI system. The training phase shows the estimation of acoustic models from individual speakers. This is usually time-consuming and hence performed offline. The evaluation phase, performed online requires fast identification of a known speaker. However, since the unknown utterance has to be compared against all enrolled speaker models, increase in the number of speakers in the set causes performance degradation (in terms of both accuracy and computational burden).

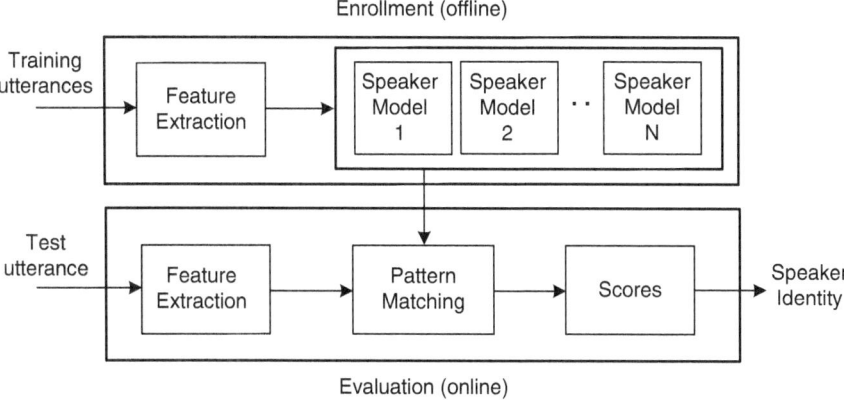

**Fig. 1.1** Block diagram of a speaker identification system

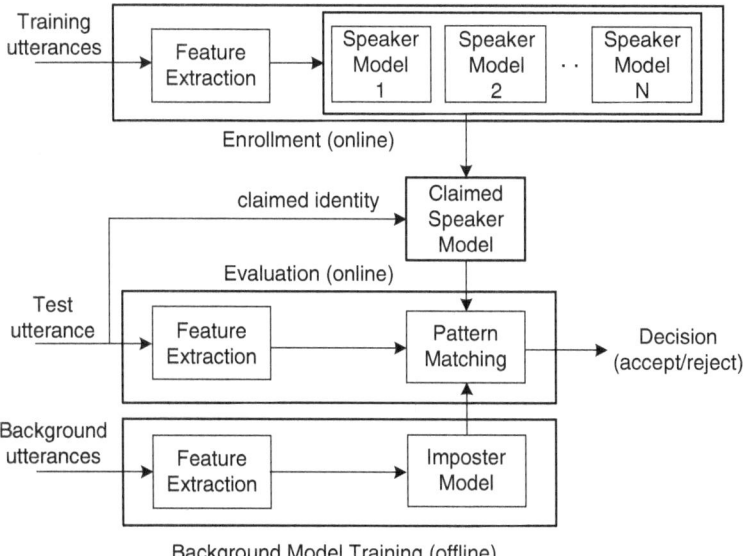

**Fig. 1.2** Block diagram of a speaker verification system

## 1.3.2 Speaker Verification

Speaker verification (SV) is the task of validating the claimed identity of a speaker. It is a binary classification problem in which the claim is either accepted or rejected based on the statistical similarities of a test utterance with the claimed speaker model (true class) and a selected background/impostor model (false class). Figure 1.2 shows the block-diagram of a typical SV system. A number of differences can be observed in contrast to SI. Firstly, a fixed pool of background speakers are required for offline training of the impostor model. The background speakers can be used as negative examples for training a discriminative model [15] or used to train a 'Universal Background Model' (UBM) [11] for GMM-based SV. In the latter case the enrolled speaker models are obtained online by adapting the UBM using a speaker's training data. Secondly, the pattern matching stage in SV requires comparison of the unknown utterance against a single claimed model and another imposter model, which makes it much faster and unaffected by the number of speakers enrolled for the SV system. The ratio of scores obtained against either model is compared with a threshold for final decision. Furthermore, SV is able to reject speech from arbitrary speakers (i.e., the open-set case) which is not true for speaker identification. Applications of ASR involving surveillance and monitoring usually require identification rather than verification. However, most online applications and security based transactions (e.g., online reservation, telebanking) require an individual to be verified rather than identified (i.e., authenticity of a claimed identity is judged irrespective of the actual identity of the speaker).

Both the above types of ASR systems may further be 'text-dependent' [19] or 'text-independent' [20]. In text-dependent systems (suitable for cooperative users) [20], the recognition phrases are fixed, or known in advance. Such systems additionally require a speech recognizer in the front-end causing more accurate but costly applications. In text-independent systems, there are no constraints on the words which the speakers are allowed to speak. Thus, the reference (what are spoken in training) and the test (what are uttered in actual use) utterances may have completely different content, and the recognition system must take this phonetic mismatch into account. Text-independent speaker recognition is thus much more challenging of the two tasks.

## 1.4 Challenging Issues in Speaker Recognition

A number of very common yet challenging issues concerning ASR, especially speaker verification has been highlighted in this section.

- **Mismatched training and test conditions:** This refers to the family of problems that arise primarily due to the differences (mismatch) in recording devices, channel, background etc., during the enrollment and evaluation phase of ASR. A typical example scenario is the development of recognition models using enrollment data acquired over the Internet and acquiring the speech data via a mobile phone during verification or testing. The medium of data acquisition or transfer seemingly encodes new information into the speech signal which largely affects the feature extraction, speaker modeling and pattern matching stages. These problems, often collectively termed as '*session variability*', has been identified as one of the most challenging issue in the field of ASR and a major source of verification errors [21,22]. The problem has been addressed over the last few decades starting with primitive methods [6] and gradually advancing into more recent techniques [21,22].

- **Intra-Speaker Variability:** While 'mismatch' occurs primarily due to extraneous factors (e.g., recording devices, background etc.), it is not solely restricted to them. Fluctuations in intrinsic/personal factors of a speaker (e.g., health, emotion, mood etc.) are also reflected across different sessions causing poor recognition. For text-independent SV systems, lack of constraints in the form of utterances spoken during training and evaluation may additionally lead to a phonetic mismatch. In general text-independent systems are more affected due to intra-speaker variability compared to text-dependent ones [23].
- **Background Noise:** Background noise is a prominent factor responsible for the loss of performance accuracy in generalized speech-based recognition tasks. Noise can be severely detrimental for ASR in both matched and mismatched conditions, the latter usually being the worse case [24]. The problem of noise or environmental degradation had been studied in past primarily in the context of speech recognition [25, 26]. A number of techniques developed for 'noise suppression' or 'noise compensation' since then, can be interchangeably applied for speaker recognition tasks. The discussion on SV for background noise shall be continued in Sect. 1.5 in more details.
- **Limited Enrollment Data:** The availability of data is a critical factor for training acoustic speaker models. The generative speaker models which are most commonly used for ASR, especially demand a high amount of training data. Usually, the required amount of training data increases proportionally with the dimension of the features extracted. This phenomenon is often termed as 'curse of dimensionality' [27]. The problem of limited data arises particularly for real-time ASR applications such as hand-held devices or in non-cooperative scenarios where speakers purposely avoid enroling for longer durations. The problem is usually tackled using statistical adaptation techniques where an already built model is modified using the acquired data [11, 28, 29].

## 1.5  Issue Addressed in Book

The book addresses the issue of speaker verification in noisy background environment. Substantial number of studies have been previously carried out in the area of robust speech recognition [25, 26]. Due to the advent of online transaction processing and the large-scale deployment of ASR technologies in hand-held devices in recent times, robustness for ASR systems has received a renewed interest [30]. In systems deployed for telephony applications the main form of degradation is due to channel variabilities induced by the handset and/or microphone. However, for speaker recognition carried out in far field applications environmental or background distortions are also of concern. As an example we may consider the typical scenario where a user enrolls for a SV system through his mobile phone while walking on a busy street. During his next access to the SV system for verification, he may be present in a secluded environment (e.g., car interior, room, office etc.). Three facts can be observed. Firstly, the background keeps changing

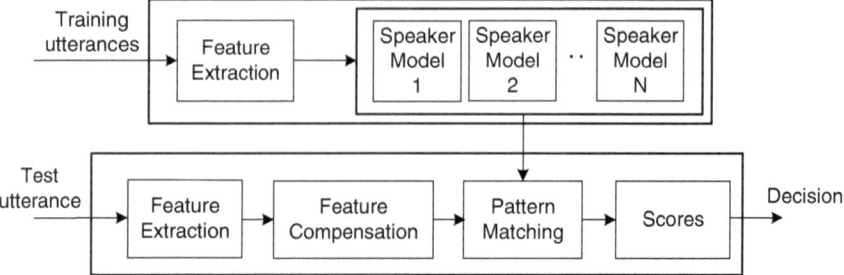

**Fig. 1.3** Block diagram of the feature compensation process

during enrollment where the user may even enter a totally unknown environment. Secondly, the obvious environmental mismatch that occurs during verification. Thirdly, there might even be a handset/channel mismatch if a different device is used during testing. In fact, in most cases especially for mismatched conditions one can expect a combined impact of both channel and background.

Background noise, in general considered additive in nature, primarily affects the spectral properties of a signal. Handling noise distortions is a challenge due to a number of reasons. Firstly, it is very difficult to quantify the effect of noise in speech primarily due to its random nature. More specifically, a clean speech segment exposed to a particular noisy environment in different intervals of time may yield noisy signals with different spectral properties. Such problems increase manifold if the noise is non-stationary i.e., its statistical properties change over time. Secondly, addition of noise results in arbitrary distortion of the feature distribution causing loss of discriminative information. This is indirectly reflected in each distinct stage of the ASR process discussed in Sect. 1.2.

The present study shall emphasize on the impact of noise in the feature-level and acoustic model-level, respectively. Noise-robustness obtained via the aforementioned stages has two broad interpretations. Firstly, the features extracted or the classifiers trained in the modeling stage may themselves be relatively immune towards the effects of channel distortions or background noise, by design. Secondly, the features and models used for generic recognition tasks in one environment may be modified or 'adapted' in another environment, to suppress the effect of mismatch. The former category comprises the group of robust features and robust speaker models while the later category comprises the family of 'compensation' or 'adaptation' techniques.

Feature compensation techniques aim to transform the features extracted during the evaluation phase such that they reflect the environmental conditions present during the training phase. Figure 1.3 shows a simplified block diagram of the feature compensation process. This is particularly applicable but not restricted to scenarios where a person enrols in a clean environment but verifies himself in a noisy one.

Despite much research for developing robust features [9], feature compensation techniques are often preferred due to the implementation costs associated with the

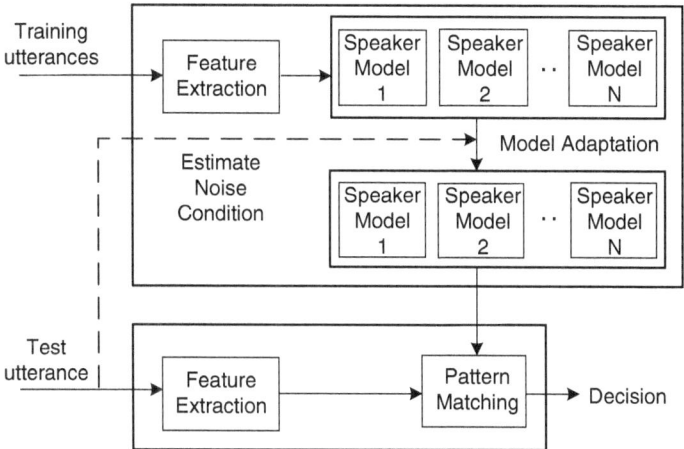

**Fig. 1.4** Block diagram of the model compensation process

former and the considerable performance improvement obtained in the latter [30]. A discussion about the various kinds of feature compensation techniques have been provided in the next chapter.

Model compensation/adaptation techniques (shown in Fig. 1.4) alters the acoustic modeling and pattern matching stages in order to account for the interfering noise. The model parameters learnt during the training phase are modified to reflect the new/mismatched environment of the evaluation phase. The traditional model compensation methods mostly rely on a priori knowledge about the test environment to adapt clean speaker models. They may be either (i) 'data-driven' in which available noisy adaptation data is used to alter pre-estimated speaker models or (ii) 'analytical' in which a mathematical structure of noise corruption is used to synthesize noisy speaker models from clean speaker models and noise models [30]. The 'data-driven' methods are usually more preferred for practical SV applications due to their low data-requirements in comparison to the 'analytical' ones which require high amount of training data. Though these methods perform significantly well (often better than feature compensation techniques), prior knowledge of test environment is sometimes considered as a major drawback for real-life scenarios. Robust speaker modeling techniques are alternatively explored as a tradeoff between accuracy and practical applications [24]. Detailed discussion about robust speaker modeling and model compensation approaches have been provided in the next chapter.

## 1.6  Objective and Scope of Work

The book aims to study alternative methods for developing ASR systems that are robust towards environmental noise. Specific focus is laid on text-independent speaker verification (SV) rather than speaker identification, since the former has a greater range of biometric applications especially in hand-held devices and online transactions.

Amongst various available strategies, the present work explores data-driven stochastic feature compensation (SFC) and robust speaker modeling methods. Two distinct categories of SFC methods based on (i) independent probability models and (ii) joint probability models, are explored. Amongst robust speaker modeling methods, the significance of supervector-based approaches in a discriminative framework for SV in noisy environment, is explored. Certain drawbacks concerning the conventional speaker modeling framework are highlighted and addressed. A boosting algorithm is proposed to combine robust discriminative classifiers for enhanced SV in degraded environments. Significance of all the methods explored in the present work is analyzed on the basis of their effectiveness and computational costs.

## 1.7  Organization of the Book

- Chapter 1 provides a brief introduction to the concept of automatic speaker recognition, its stages, categories and modern applications. A number of challenging issues in the field of ASR are highlighted. A brief discussion of the issue addressed in the book is provided followed by the objective and scope of work.
- Chapter 2 provides an overview of various feature and model-based approaches developed in past for robust speaker recognition. The advantages and disadvantages of some standard methods applied for robust SV tasks have been highlighted.
- Chapter 3 discusses baseline SV systems developed using the GMM-UBM framework in noisy environments. A feature mapping technique using multiple background model framework has been explored for robust SV in time-varying noisy environments.
- Chapter 4 explores the impact of standard stereo-based stochastic feature compensation (SFC) methods for robust speaker verification in uniform noisy environments. Integration of a SFC stage in the GMM-UBM framework is proposed for SV evaluation under mismatched conditions.
- Chapter 5 explores robust speaker-modeling methods for SV in noisy environments. Specifically, the combined GMM-SVM and SVM-i vector approaches are used for developing SV systems and evaluating them in matched conditions
- Chapter 6 provides a brief summary and conclusion of the Book.

## 1.8 Contribution of the Book

The contribution of the book lies in exploring feature compensation and robust speaker modeling methods, the impact of which have not been erstwhile studied explicitly for speaker verification in noisy environments. The major contributions can be broadly summarized under the following points

- A class of data-driven stochastic feature compensation methods has been explored for robust speaker verification (SV) in noisy background environments.
- The robustness of some state-of-the-art speaker modeling methods (e.g., GMM supervector, i-vector) in a discriminative framework using SVM classifiers, has been explored for SV in noisy environments.
- A novel boosting algorithm is proposed for combining robust SVM classifiers for improving SV performance.

## References

1. J.P. Campbell, W. Shen, W.M. Campbell, R. Schwartz, J.F. Bonastre, D. Matrouf, Forensic speaker recognition. IEEE Signal Process. Mag. **26**(2), 95–103 (2009)
2. B.G.B. Fauve, D. Matrouf, N. Scheffer, J.F. Bonastre, J.S.D. Mason, State-of-the-art performance in text-independent speaker verification through open-source software. IEEE Trans. Audio Speech Lang. Process. **15**(7), 1960–1968 (2007)
3. T. Kinnunen, H. Li, An overview of text-independent speaker recognition: from features to supervectors. Speech Commun. **52**, 12–40 (2010)
4. D.V. Lancker, J. Kreiman, K. Emmorey, Familiar voice recognition: patterns and parameters Part I: recognition of backward voices. J. Phon. **13**, 19–38 (1985)
5. A. Reich, J. Duke, Effects of selected vocal disguises on speaker identification by listening. J. Acoust. Soc. Am. **66**(4), 1023–1029 (1979)
6. G. Doddington, Speaker recognition—identifying people by their voices. Proc. IEEE **73**(11), 1651–1664 (1985)
7. J. Wolf, Efficient acoustic parameters for speaker recognition J. Acoust. Soc. Am. **6**(51), 2044–2056 (1972)
8. S. Davis, P. Mermelstein, Comparison of parametric representations for monosyllabic word recognition in continuously spoken sentences. IEEE Trans. Acoust. Speech Signal Process. **28**(4), 357–366 (1980)
9. D.A. Reynolds, Experimental evaluation of features for robust speaker identification. IEEE Trans. Speech Audio Process. **2**(4), 639–643 (1994)
10. D.A. Reynolds, R.C. Rose, Robust text-independent speaker identification using Gaussian mixture speaker models. IEEE Trans. Acoust. Speech Signal Process. **3**(1), 72–83 (1995)
11. D. Reynolds, T. Quatieri, R. Dunn, Speaker verification using adapted Gaussian mixture models. Digit. Signal Process. **10**(1), 19–41 (2000)
12. M. BenZeghiba, H. Bourland, On the combination of speech and speaker recognition, in *Proceedings of the European Conference of Speech Communication and Technology (EUROSPEECH '03)*, Geneva, 2003
13. D. Burton, Text-dependent speaker verification using vector quantization source coding. IEEE Trans. Acoust. Speech Signal Process. **35**(2), 133–143 (1987)
14. L. Rabiner, B.H. Juang, *Fundamentals of Speech Recognition*, 1st edn. (Prentice-Hall, Englewood Cliffs, 1993)

15. W. Campbell, J. Campbell, D. Reynolds, E. Singer, P. Carrasquillo, Support vector machines for speaker and language recognition. Comput. Speech Lang. **20**, 210–229 (2006)
16. K. Farrell, R. Mammone, K. Assaleh, Speaker recognition using neural networks and conventional classifiers. IEEE Trans. Speech Audio Process. **2**(1), 195–204 (1994)
17. W. Campbell, J. Campbell, D. Reynolds, Support vector machines using GMM supervectors for speaker verification. IEEE Signal Process. Lett. **13**(5), 308–311 (2006)
18. B. Yegnanarayana, S.P. Kishore, AANN: an alternative to GMM for pattern recognition. Neural Netw. **15**, 456–469 (2002)
19. F. Bimbot, J.F. Bonastre, C. Fredouille, G. Gravier, I. Magrin-Chagnolleau, S. Meignier, T. Merlin, J. Ortega-Garcia, D. Petrovska-Delacrétaz, D.A. Reynolds, A tutorial on text-independent speaker verification. EURASIP J. Adv. Signal Process. (Spec. Issue Biom. Signal Process.) **4**(4), 430–451 (2004)
20. M. Hebert, Text-dependent speaker recognition, in *Springer Handbook of Speech Processing* (Springer, Berlin/Hiedelberg, 2008), pp. 743–762
21. P. Kenny, G. Boulianne, P. Ouellet, P. Dumouchel, Speaker and session variability in GMM-based speaker verification. IEEE Trans. Audio Speech Lang. Process. **15**(4), 1448–1460 (2007)
22. R. Vogt, S. Sridharan, Explicit modeling of session variability for speaker verification. Speech Commun. **1**(22), 17–38 (2008)
23. J. Kahn, N. Audibert, S. Rossato, J.F. Bonastre, Intra-speaker variability effects on speaker verification performance, in *The Speaker and Language Recognition Workshop (Odyssey '10)*, Brno, 2010
24. J. Ming, T.J. Hazen, J.R. Glass, D. Reynolds, Robust speaker recognition in noisy conditions. IEEE Trans. Audio Speech Lang. Process. **15**(5), 1711–1723 (2007)
25. A. Acero, Acoustical and environmental robustness in automatic speech recognition. PhD thesis, Carnegie Mellon University, Sept 1990
26. P. Moreno, Speech recognition in noisy environments. PhD thesis, Electrical & Computer Engineering Department, Carnegie Mellon University, Pittsburgh, 1996
27. C.M. Bishop, *Pattern Recognition and Machine Learning* (Springer, New York, 2006)
28. J. Gauvain, C. Lee, Maximum a posteriori estimation for multivariate Gaussian mixture observations of Markov chains. IEEE Trans. Speech Audio Process. **2**(2), 291–298 (1994)
29. V. Hautamaki, T. Kinnunen, I. Karkkainen, M. Tuononen, J. Saastamoinen, P. Franti, Maximum a posteriori adaptation of the centroid model for speaker verification. IEEE Signal Process. Lett. **15**, 162–165 (2008)
30. R. Togneri, D. Pullella, An overview of speaker identification: accuracy and robustness issues. IEEE Circuits Syst. Mag. **11**(2), 23–61 (2011)

# Chapter 2
# Robust Speaker Verification: A Review

**Abstract** This chapter provides an overview of various feature and model-based approaches developed in past for robust speaker recognition. The advantages and disadvantages of some standard methods applied for robust speaker verification tasks have been highlighted. The main focus is to summarily introduce popular state-of-the-art techniques adopted for enhancing speaker verification performance in noisy conditions.

This chapter provides a broad overview of research methods developed for robust speaker recognition tasks in past. The focus is to summarily introduce popular state-of-the-art techniques adopted for enhancing speaker verification performance in noisy conditions, especially those within the current scope of work. The chapter mainly emphasizes in the feature extraction and statistical modeling stages of speaker recognition. The merits and de-merits of some of these techniques are discussed in the purview of the book. It is to be noted that many of these methods have primarily been applied for robust speech recognition in noisy environment. Since some of the intermediate stages of speaker verification are similar to that of speech recognition, they may be interchangeably used for the former. The readers are encouraged to follow the references for detailed description of the methods discussed especially notable reviews such as [1–4] or recent research works [5]. Concise overviews of methods adopted for feature compensation, feature extraction, model compensation and robust speaker modeling are briefly presented in different sections of this chapter. The role of each of these stages has been discussed in the first chapter. The final section briefly describes the motivation of carrying out the present research work.

K.S. Rao and S. Sarkar, *Robust Speaker Recognition in Noisy Environments*,
SpringerBriefs in Electrical and Computer Engineering,
DOI 10.1007/978-3-319-07130-5_2, © The Author(s) 2014

## 2.1   Feature Compensation

Ever since parameterization of raw speech signal was first studied [6], the motivation was to discover speaker-discriminative features for generalized recognition tasks [7]. The significance of cepstral features [8], especially the mel-cepstrum [9] for speaker recognition (SR) was established during the contemporary period. However, there were practical limitations of the use of cepstral features due to arbitrary modification of the cepstral distribution in the presence of channel distortions or background noise. A series of feature compensation techniques were proposed during the early 1990s as a refinement of the common feature extraction process [10–12]. The motivation was to make real-life applications of SR or speech recognition which countered channel-induced distortions and handset mismatches over telephonic conversations [13]. The class of feature compensation methods developed since then, may be broadly categorized into three groups i.e., filtering-based compensation, noise model-based compensation and empirical compensation. Apart from the conventional compensation techniques, there exists a group of feature transformation methods which are often used in conjunction with the former. In [14], neural network models are used as mapping functions for transforming the emotion-specific features to emotion-independent features for developing robust speaker recognition system.

The filtering techniques aim to denoise or suppress the effect of noise in the extracted features. They exploit the fact that convolutive channel or environmental distortions become additive in the log-spectral and cepstral domain. It was studied in [15] that slow variations in the channel appear as an offset of individual co-efficients of a cepstral vector. Cepstral Mean Subtraction (CMS) [15] suppresses the channel effects by subtracting the mean of cepstral co-efficients extracted from short-term frames, from the individual coefficients. The removal of the average spectrum also suppress inter-session variabilities to certain extent [11]. Apart from simple mean-removal as in CMS the variance of the cepstral vectors are often scaled to unity. Relative Spectra (RASTA) [16], principally similar to CMS, was proposed to compensate for rapidly varying channel conditions. Instead of uniform mean subtraction over the entire cepstra, a moving average filter was employed for an exponentially decaying mean subtraction. CMS and RASTA are commonly applied for front-end compensation in SR tasks due to the simplicity of implementation. A set of more sophisticated '*kernel filtering*' methods [17] were later developed which captured the non-linear features of speech by fitting a higher dimensional mapping function and eventually projecting the features to a lower dimensional manifold. However later studies had promptly revealed that these techniques are not much effective for channel mismatches and additive background noise.

The model-based feature compensation methods assume a priori knowledge of the noise spectrum. An estimate of the clean speech parameters is made using either a noise model or representation of the effects of noise in speech. The primitive methods in this group include Spectral Equalization [18] and Spectral Subtraction (SS) [19]. In SS, the clean speech spectra is estimated by subtracting the mean magnitude of an approximate noise spectra from that of the noisy speech spectra. These methods relied on the stationary assumption of noise and independence of spectral estimates across frequencies explicitly. To overcome this limitation, some of the methods developed later were based on the minimum mean squared error (MMSE) predictor [20] which modeled the correlation of frequency components e.g., MMSE log spectral amplitude estimator [21]. During the early 1990s, stereo-data based compensation techniques were first introduced [10]. Cepstral compensation vectors were derived from a stereo database and applied to the training data to adapt to environmental changes. The compensation could also be in the form of affine transformations learned from stereo data [12]. Popular examples are Codeword Dependent Cepstral Normalization (CDCN) [22] and its variants like Fast CDCN (FCDCN) [10]. Other methods relied on a mathematical model of the environmental mismatch due to noise. The parameters of the model were estimated and applied to the appropriate inverse operation to compensate the test signal e.g., feature-level Vector Taylor Series [23].

The third group of feature compensation techniques are entirely data-driven and are stochastic in nature. They are 'blind' towards the nature of the corrupting process and are based on empirical compensation methods that use direct spectral comparison. Prior work shows that they often outperform the previous two approaches for feature enhancement [24]. During the training phase, some transformations are estimated by computing the frame-by-frame differences between the vectors representing speech in the clean and noisy environments (stereo data). The differences between clean and noisy feature vectors are modeled by training additive bias vectors on the mean and covariance of either of the two (clean or noisy) probability distributions. During evaluation phase, the bias vectors are used to transform noisy test feature vectors to their clean feature equivalent based on the MMSE predictor. Previous MMSE-based methods like CDCN [22] and FCDCN [10], used vector quantization (VQ) codebooks to represent the distribution of clean feature vectors. Due to their quantization-based framework, these algorithms were unable to learn the variance of a distribution and were later replaced by the more flexible Gaussian Mixture Model (GMM)-based normalization techniques e.g., Multivariate Gaussian-based Cepstral Normalization (RATZ) [25]. Although the RATZ family of algorithms approximated the normalized features, the posterior probability of clean GMM components with respect to the noisy test feature vectors were usually distorted, causing poor MMSE estimates. To suppress these distortions, the Stereo-based Piecewise Linear CompEnsation for Environments (SPLICE) algorithm proposed in [26] modeled the noisy feature space using GMMs instead. This produced significantly better result in robust speech recognition tasks compared to its predecessors [27]. The effectiveness of SPLICE framework has since then encouraged its extended applications e.g., speech recognition in non-stationary noisy environments within cars using the Multi Environment Model-based Linear Normalization (MEMLIN) algorithm [28] and word recognition using Noise Adaptive Training [27]. The more recently proposed Stereo-based Stochastic Mapping (SSM) [29] is principally a more accurate version of SPLICE based on joint probability modeling of the noisy and clean feature spaces using GMMs.

## 2.2  Robust Feature Extraction

The conventional features used for SR tasks can be broadly categorized as spectral, prosodic and high-level features. In this section we briefly discuss each.

Prosody is a collective term for certain aspects manifested in long term speech segments e.g., stress, intonation pattern, rhythm etc. The most significant amongst these is intonation which is characterized by the fundamental frequency contour ($F_o$). $F_o$ contour and energy (stress) were effectively used for speaker recognition in [30]. A few other significant applications of prosodic features for SR include combination of energy trajectory with $F_o$ [31] and construction of SVM speaker models using pitch, duration and pause features [32]. In [33], temporal variations in speaker-specific prosodic parameters are proposed in addition to conventional spectral features for improving the speaker recognition accuracy in presence of noisy background environments. A comparative study about the significance of the various prosodic features for SR tasks can be found in [34]. Modeling the different levels of prosodic information (instantaneous, long-term) for speaker discrimination is considered to be a difficult task. At the same time, it is desired that the features are free from the effects that a speaker can voluntarily control. Due to these complications, prosodic features haven't been much used for robust SR tasks.

High-level features exploit speaker's choice of words or vocabulary for recognizing them. The term 'high-level' refers to modeling speech utterances using the sequence of 'tokens' present in them. The co-occurrence pattern in the tokens, often termed as 'idiolect' [35], characterizes speaker differences. The tokens that are commonly used for speaker recognition may be in the form of phones [36], words [35], prosodic gestures [31, 34] or even articulatory movements [37]. Significant applications of these features for SR include [36, 38], where GMMs trained using individual sets of extracted tokens are used in parallel for classification. Due to their nature, high-level features can often be interchangebly used for speaker and language recognition [39]. Other approaches share similarities with the common prosodic features [31]. A study on the joint application of prosodic and high-level features for robust SR tasks can be found in [40]. However, high-level features are not a very attractive group to work with, due to the computational complexity involved in recognizing tokens.

The most common features for generalized speech related tasks as well as speaker recognition, are the family of spectral or spectro-temporal features. These features are extracted from short overlapping frames (10–25 ms) which are pre-emphasized and smoothed. Based on their interpretation they can be categorized as temporal, spectral or spectro-temporal. Popular examples of cepstral features are the Linear Prediction Cepstral Coefficient (LPCC) [8], Perceptual Linear Prediction (PLP) [41] coefficients and Mel Frequency Cepstral Coefficient (MFCC) [9], etc.

LPCCs are based on the principle of correlation of a sample with its adjacent ones. An instantaneous sample is approximated in terms of its neighborhood samples weighted with a set of predictor coefficients. The error in estimation is often termed as LP residual. The frequency domain equivalent of this representation is that of an all-pole filter with the same set of LP coefficients. The coefficients are determined by minimizing the residual energy using the Levinson Durbin algorithm [42]. The prediction coefficients instead of being used by themselves are transformed into a set of robust, less correlated features like LPCCs, PLP [41], Line Spectral Frequencies (LSF) [43], formant frequencies and bandwidth etc. [42].

The MFCCs [9] are the most successful and extensively used features for speaker recognition. MFCCs were psychoacoustically motivated in the sense that they were found to mimic the human auditory perception. MFCCs are extracted by a non-linear filter-bank analysis of the Discrete Fourier Transform (DFT) magnitude spectrum of short-term frames. The filterbank usually consist of a set of triangular band-pass filters, which are spaced according to the 'mel' scale. The log-magnitude of the filtered spectra is subjected to a Discrete Cosine Transform (DCT) for obtaining the cepstral features. MFCCs have arguably shown the best results compared to contemporary features like LPCC, PLP, LSF etc., in several prior works in SR using clean speech [10–12, 44]. Thus they are considered to be the default features for several speech related tasks including SR.

However the presence of background noise or channel effects inhibit the performance of MFCCs significantly primarily due to the distortion in the feature distribution [25]. The default use of MFCCs in most baseline SR systems necessitated the development of feature compensation methods as discussed in Sect. 2.1. However quite recently, researchers have focussed on alternative ways of modifying the cepstral feature extraction process for resistance towards ambient noise. Amongst several others, some notable features are Mean Hilbert Envelope Coefficient (MHEC) [45], Power Normalized Cepstral Coefficient (PNCC) [46] and Normalized Modulation Cepstral Coefficient (NMCC) [47]. Instead of modifying the features for compensating the effect of noise, features can be extracted from selective-regions of speech. Even in presence of noise also, glottal closure region in each pitch cycle and steady vowel regions contain high signal to noise ratio, and hence, features extracted from these regions are more robust compared to other regions of speech. In [48–50], features extracted from above mentioned regions are explored for robust speaker and language recognition tasks.

In MHEC extraction, the pre-emphasized speech is first decomposed into a number of spectral subbands using a gammatone filter constrained in the telephonic bandwidth of 300–3400 Hz. Unlike MFCC, the filters are uniformly spaced on an equivalent rectangular bandwidth (ERB) scale. The temporal envelope (Hilbert envelope) of each subband is estimated by using the Hilbert transform of the subband signal followed by low-pass filtering. The smoothed envelope is then used for deriving the required cepstral features. In PNCC extraction, the pre-emphasized signal is analyzed using short overlapping frames. A short-time Fourier analysis is performed over the Hamming windowed data, followed by frequency domain filtering using a gammatone filterbank constrained in 133 and 4,000 Hz, where the center frequencies of the gammatone bank are spaced equally in the ERB scale. The NMCCs are similar to PNCC except that amplitude modulation (AM) signals are estimated from the gammatone filtered sub-band signals using a Teager non-linear energy operator. The resulting signal is power normalized followed by DCT transform to obtain the cepstral features.

## 2.3  Model Compensation

Though feature-level compensation techniques are often applied as a front-end denoising process due to their low computational complexity and independence of any recognition model, they have certain limitations. In most cases, the feature compensation techniques produce point estimates of clean speech features. Due to this, they are unable to capture the uncertainty of observations which is represented as the variance of the conditional distribution of noisy speech given clean speech [51]. An alternative is to alter the statistical parameters of the acoustic model learned during the training phase to compensate for the channel or environmental mismatch of the evaluation phase. Since the evolution of statistical models for speech recognition, much research has been devoted in exploring model compensation issues in parallel [52, 53].

The earlier methods focussed on rendering the speaker models ineffective towards channel mismatches or handset variations [54]. In most cases, the mismatch would be caused due to unseen channel data during the evaluation phase. Unlike speech recognition tasks where multiple channel adaptation data could be obtained by pooling all speaker data over individual channels, SR required speaker-specific enrollment speech over multiple channels which could be later used for verification. This was unfavourable for practical SR applications. Alternate methods would cluster the data from a single conversation into multiple channel types to meet data requirements. Synthetic variance distribution [55] used an auxillary database of stereo recordings to artificially construct a global distribution of variances. Transformations derived from this distribution were used to modify the variance of individual speaker models. Speaker Model Synthesis (SMS) [56] learned speaker-independent transformations between different channels and applied it to synthesize speaker models under unseen enrollment conditions. The transformations were learned in the form of mean shift, weight scaling and variance scaling of GMM model parameters trained across various channel conditions.

In contrast to model-based channel compensation schemes, model-based environment adaptation methods developed during the contemporary period, modify speaker model parameters to reflect the acoustic environment of the evaluation phase. Two most popular data-driven environmental adaptation techniques initially proposed for robust speech recognition are *Maximum aPosteriori* (MAP)[57] and Maximum Likelihood Linear Regression (MLLR) [58]. The successful application of GMMs in the field of speaker recognition [59] has since then encouraged their usage in robust speaker verification (SV) tasks [60]. Both these methods use adaptation data to build speaker-specific models from a speaker independent background model constructed offline. MAP is a two stage process in which Bayesian statistics estimated using the training/adaptation data in the first stage, are used to update the 'a priori' available background model parameters (mean, covariances and weights) in the second stage. The Speaker Model Synthesis [56] method was based on deriving individual channel dependent GMMs by MAP-adaptation of a channel-independent background model. In another application,

MAP was jointly used for model adaptation as well as feature transformation [61]. The advantage of MAP adaptation is its close approximation to the ideal maximum-likelihood estimates given sufficient enrollment data. However in situations where training data is sparse, MAP would only update a fractional number of GMM components. The MLLR adaptation technique transform the background GMM means and covariance matrices (optionally) by an affine transformation aiming at maximizing the likelihood function given new adaptation data. The parameters of the transformation are derived by iteratively using the Expectation Maximization (EM) algorithm [62]. Unlike MAP, all the GMM components are updated with limited amount of enrollment data. Other variants of MLLR like constrained MLLR (CMLLR) [63] are often used for online model adaptation [64]. However the performance improvement in MLLR-based methods saturates with increasing adaptation data and at a certain stage they are outperformed by MAP. A comparison of MLLR and Neural Network based environmental techniques was made in [65].

Apart from the traditional data driven methods that are dependent on adaptation data representing acoustic conditions of the evaluation phase, another approach is to exploit a priori information about the test environment. Popular state-of-the-art techniques in this category are Parallel Model Combination (PMC) [66, 67] and Vector Taylor Series (VTS) [68]. PMC relies on an available statistical noise model of the recognition phase and clean speaker GMMs trained during enrollment. The aim is to obtain noise-corrupted model for pattern matching, by combining the clean speech and noise models. This is done in two stages. Firstly, clean speaker models (GMMs/HMMs) and a simplified noise model (GMM) are built independently from clean training data and a noise signal, respectively. Secondly, the effect of additive noise on clean speech in the cepstral domain is analysed by using a function of noise corruption. This function is then extended to the parametric space to estimate the corrupted model parameters (mean and variances) from the clean and noise model parameters, respectively. Prior work shows that PMC model parameter estimation gets increasingly complex for dynamic and acceleration coefficients of MFCC. A recent state-of-the-art technique [69] addresses this problem by exploiting the relation between static and dynamic coefficients. The VTS method [70] uses a similar mathematical structure to represent the noise corruption process. However, unlike PMC the noise and channel statistics are obtained via an approximate taylor series expansion of the function around the mean of GMM components. This method is relatively much simpler compared to PMC and the tradeoff in terms of accuracy is not significant.

Though the model-compensation techniques perform better than their feature-level counterparts, they are computationally intensive and often require substantial amount of training data. Apart from the two broad types of compensation techniques discussed in Sects. 2.1 and 2.3, there exists hybrid approaches which can be termed as a combination of the two methods. Examples include Stochastic Matching [71] and Joint Uncertainty Decoding [72]. These methods account for the imperfections in feature enhancement process by approximating the marginal distribution of noisy features. In realistic situations, it may also turn out that the verification environment is entirely unknown [73]. In such scenarios, one might not expect availability of adaptation data or stereo training data. Quite recently, researchers have addressed this issue [74] by combining 'missing feature theory' based techniques [75] to subdue noise variation outside training conditions. The 'posterior union model' in [74], require detection and exclusion of the heavily mismatched subbands of the speech spectra. However, the improvement in performance accuracy of all these methods is usually associated with increased computational load and dependency on numerical approximations.

## 2.4 Robust Speaker Modeling

Speaker modeling techniques have been extensively explored in the past few decades of SR research. The scope of applying diverse pattern recognition techniques for classification and clustering of features makes this field an exciting area to work with. The broad classes of modeling techniques that are used in practice can be broadly categorized as generative models (GMM, VQ, Joint Factor Analysis (JFA)) or discriminative models (Neural Networks (NN) and Support Vector Machines (SVMs)). A family of hybrid modeling techniques also exist which are a combination of both e.g., GMM-SVM, SVM-JFA, etc. In this subsection we shall briefly discuss each.

Vector Quantization, introduced in the late 1980s [76] is one the most primitive form of SR model. Based on the principle of K-means [62], the set of feature vectors extracted from a speaker's training utterance are grouped into a number of non-overlapping clusters. Individual speaker models are represented by the stack of cluster centroids often termed as codebook. Classification of a test utterance is based on minimization of a distortion measure commonly given by the average Euclidean distance of a vector from each codebook. Despite its crude form of clustering, VQ is often used for computational speedup required for real-time SR applications [77].

The Gaussian Mixture Models (GMMs) introduced in the mid-1990s [59] is widely considered to be a benchmark for modern text-independent Speaker Recognition. In contrast to VQ, a number of overlapping multivariate Gaussian functions are used to cluster the feature space. The GMMs are able to characterize general properties like multi-modal feature distribution, speaker-dependent spectral shapes etc. Unlike VQ, GMMs are able to capture the variance of feature distribution. In contrast to the naive K-means, GMM training is based on a more rigorous approach of maximizing the likelihood of a given speaker's data. The parameters are estimated iteratively using the Expectation Maximization (EM) algorithm [62]. Classification of test utterances are done on the basis of log likelihood scores obtained from the sequence of test vectors. Though speaker-specific GMMs performed reasonably well for SR given clean speech, a good amount of data was required for parameter estimation. Besides, a more generalized approach was required for unifying model-compensation techniques with the GMM framework. A novel GMM-based approach was proposed in [60], where a single speaker-independent GMM (Universal Background Model (UBM)) trained using multiple speaker data across various channels and sessions, was used as a common impostor model for speaker verification. HMM-based speaker models were derived using MAP and MLLR adaptation of the UBM using the speaker's training data. Besides reducing data requirements, these techniques provided scope for model adaptation as discussed in Sect. 2.3. Comparative studies of alternate adaptation techniques were made in [78]. Efforts were also made to approximate the common MAP adaptation process in terms of a VQ model [79]. However in the context of environmental robustness, GMMs often provide limited performance improvement despite model-adaptation. This problem received a new direction with the introduction of Joint Factor Analysis and its variants [80].

Prior to the introduction of GMMs, role of Neural Networks (NN) for text-independent SR was first studied in [81]. An advantage of NNs is its ability to perform feature transformation and speaker modeling simultaneously [82]. In a later study, Auto Associative Neural Networks (AANNs) were introduced for speaker modeling [83]. Since GMMs relied on first and second order statistics, it was hypothesized in [83] that they fail to capture feature distribution based on higher order statistics. AANNs were found to be effective for SR tasks where distribution of data is highly non-linear [83]. However, NNs have not been used much in practice primarily due to the heavy computational costs involved in training them. Besides, prior determination of the appropriate structure for NNs (number of neurons in each layer) is a non-trivial task.

Support Vector Machines (SVMs) have emerged as a powerful discriminative classifier in the field of robust SR in the last decade [32, 36, 84]. A SVM is a binary classifier which distinguishes between two classes (true speaker and impostor) by learning a decision hyperplane which separates them in some higher dimensional feature space [62]. SVMs have been initially used to model individual speakers using high-level [36] and prosodic features [32]. However the real significance of SVMs in robust SV tasks was found in its effective combination with the traditional GMM classifier [85]. A novel method of representing variable length training utterances using fixed-length vectors was discovered contemporarily. The mean vectors of MAP-adapted speaker GMMs were stacked together to produce a high dimensional vector commonly termed as a 'supervector'. The labelled supervectors were used as input the SVMs. This led to the scope of exploring various 'sequence kernels' or non-linear mappings for transforming features to high dimensional spaces [85–87]. Several normalization techniques for minimizing inter-session and intra-speaker variabilities in the supervector space have been introduced since then. Common examples are Nuisance Attribute Projection (NAP) [88], Within Class Covariance Normalization (WCCN) [89] and Linear Discriminant Analysis (LDA) [62]. The GMM-SVM approach is often considered as a effective alternative of the GMM-UBM method.

Supervector-based speaker recognition opened an interesting new direction for compensating channel and session variabilities. It was thought that channel variations in recorded training utterances might lead to the problem of mismatch in the supervector space. A feasible alternative was to explicitly model the channel variability by representing the supervector space as a combination of statistically independent channel and speaker subspaces. This approach was named Joint Factor Analysis (JFA) [80] where the term 'factor' denotes the low-dimensional projection of the speaker or channel supervectors in their corresponding spaces. JFA as a new research trend has been extensively studied for robust SR tasks since the late 2000s [80, 90]. However it was later argued in [91], that instead of two distinct subspaces a single 'total variability' space could in fact be useful for simultaneously representing both speaker and channel variabilities. A low-dimensional projection of the supervectors in the total variability space, commonly known as 'i-vectors' has since then been considered as the modern state-of-the-art in robust speaker verification. Various studies have since then been conducted to combine JFA and SVM based methods with appropriate normalization techniques [92]. Quite recently, i-vector based studies have conducted for robust speaker recognition tasks where authors have proposed alternative methods of projecting the i-vectors into a subspace for improved speaker discrimination and suppression of channel-effects [93].

## 2.5   Motivation for the Present Work

Robust speaker recognition in noisy environments till date remains an open issue despite the diverse array of methods developed to address it in past. The ever increasing usage of hand-held devices in the modern era has driven new demand for robust speaker recognition applications. Despite being well explored in past, new methods keep unfolding in this field which are either suggested improvements or alternatives of the existing ones. This makes robust SR a very challenging and yet an interesting area to work in.

Despite the availability of robust features as discussed in Sect. 2.2, feature compensation techniques play a crucial role for SV applications that demand noise-robustness without compromising on speaker-discriminative power [94]. An interesting fact to notice about the state-of-the art data-driven feature compensation methods discussed in Sect. 2.1, is that their application has mostly been restricted to robust speech recognition tasks but rarely studied for robust SV tasks. The brief discussion about model compensation techniques in Sect. 2.3 reveal some of their vulnerabilities. They either rely explicitly on an available clean speaker model (e.g., PMC [66], VTS [70]) or a priori knowledge about the noisy environment (e.g., noise model for PMC, adaptation data for MAP [57], MLLR [58]). These drawbacks suggest the use of robust speaker modeling methods as an alternative for practical scenarios (e.g., unknown noisy environment, unavailable clean speaker models). In a similar context it can be argued that the state-of-the-art robust speaker modeling methods (e.g., GMM supervectors [85], i-vectors [95] etc.) have mostly been applied to counter channel/handset mismatches but not additive background noise specifically.

Summarily, the above two points motivates us to propose new studies in which we explore the application of feature enhancement techniques for speaker verification in additive background noise. Studies are also conducted to demonstrate the effectiveness of supervector-based approaches and its state-of-the-art variants (e.g., i-vectors) for robust speaker verification in noisy environments.

# References

1. J. Campbell, Speaker recognition: a tutorial. Proc. IEEE **85**(9), 1437–1462 (1997)
2. F. Bimbot, J.F. Bonastre, C. Fredouille, G. Gravier, I. Magrin-Chagnolleau, S. Meignier, T. Merlin, J. Ortega-Garcia, D. Petrovska-Delacrétaz, D.A. Reynolds, A tutorial on text-independent speaker verification. EURASIP J. Adv. Signal Process. (Spec. Issue Biom. Signal Process.) **4**(4), 430–451 (2004)
3. B.G.B. Fauve, D. Matrouf, N. Scheffer, J.F. Bonastre, J.S.D. Mason, State-of-the-art performance in text-independent speaker verification through open-source software. IEEE Trans. Audio Speech Lang. Process. **15**(7), 1960–1968 (2007)
4. T. Kinnunen, H. Li, An overview of text-independent speaker recognition: from features to supervectors. Speech Commun. **52**, 12–40 (2010)
5. S. Sarkar, Robust speaker recognition in noisy environments. Master's thesis, School of Information Technology, Indian Institute of Technology Kharagpur, Mar 2014
6. R. Schafer, L. Rabiner, Digital representations of speech signals. Proc. IEEE **63**(4), 662–677 (1975)
7. B. Atal, Automatic recognition of speakers from their voices. Proc. IEEE **64**(4), 460–475 (1976)
8. J. Makhoul, Linear prediction: a tutorial review. Proc. IEEE **63**(4), 561–580 (1975)
9. S. Davis, P. Mermelstein, Comparison of parametric representations for monosyllabic word recognition in continuously spoken sentences. IEEE Trans. Acoust. Speech Signal Process. **28**(4), 357–366 (1980)
10. A. Acero, Acoustical and environmental robustness in automatic speech recognition. PhD thesis, Carnegie Mellon University, Sept 1990
11. D.A. Reynolds, Experimental evaluation of features for robust speaker identification. IEEE Trans. Speech Audio Process. **2**(4), 639–643 (1994)
12. R. Mammone, X. Zhang, R. Ramachandran, Robust speaker recognition: a feature-based approach. IEEE Signal Process. Mag. **13**(5), 58–71 (1996)
13. D. Reynolds, The effects of handset variability on speaker recognition performance: experiments on the Switchboard corpus, in *Proceedings of IEEE International Conference on Acoustics, Speech and Signal Processing*, Atlanta, 1996, vol. 1, pp. 113–116
14. K.S. Rao, J. Yadav, S. Sarkar, S.G. Koolagudi, A.K. Vuppala, Neural network based feature transformation for emotion independent speaker identification. Int. J. Speech Technol. (Springer) **15**(3), 335–349 (2012)
15. S. Furui, Cepstral analysis technique for automatic speaker verification. IEEE Trans. Acoust. Speech Signal Process. **29**(2), 254–272 (1981)
16. H. Hermansky, N. Morgan, RASTA processing of speech. IEEE Trans. Speech Audio Process. **2**(4), 578–589 (1994)
17. A. Kocsor, L. Toth, Kernel-based feature extraction with a speech technology application. IEEE Trans. Signal Process. **52**(8), 2250–2263 (2004)
18. T.G. Stockham, T.M. Cannon, R.B. Ingebretsen, Blind deconvolution through digital signal processing. Proc. IEEE **63**(4), 678–692 (1975)
19. S. Boll, Suppression of acoustic noise in speech using spectral subtraction. IEEE Trans. Acoust. Speech Signal Process. **27**(2), 113–120 (1979)
20. A. Erell, M. Weintraub, Spectral estimation for noise robust speech recognition, in *Proceedings of DARPA Speech and Natural Language Workshop*, Philadelphia, 1989

21. Y. Ephraim, D. Malah, Speech enhancement using a minimum mean-square error log-spectral amplitude estimator. IEEE Trans. Acoust. Speech Signal Process. **33**(2), 443–445 (1985)
22. A. Acero, R.M. Stern, Environmental robustness in automatic speech recognition, in *Proceedings of IEEE International Conference on Acoustics, Speech and Signal Processing (ICASSP '90)*, Albuquerque, 1990, vol. 2, pp. 849–852
23. S. Suhadi, S. Stan, T. Fingscheidt, C. Beaugeant, An evaluation of VTS and IMM for speaker verification in noise, in *Proceedings of 4th Annual Conference of the International Speech Communication Association (INTERSPEECH '03)*, Geneva, 2003, pp. 1669–1672
24. L. Deng, J. Droppo, A. Acero, Recursive estimation of non-stationary noise using iterative stochastic approximation for robust speech recognition. IEEE Trans. Speech Audio Process. **11**(6), 568–580 (2003)
25. P.J. Moreno, B. Raj, R.M. Stern, Data-driven environmental compensation for speech recognition: a unified approach. Speech Commun. **24**(4), 267–285 (1998)
26. L. Deng, A. Acero, M. Plumpe, X. Huang, Large-vocabulary speech recognition under adverse acoustic environments, in *Proceedings of the International Conference of Spoken Language Processing (ICSLP '00)*, Beijing, 2000, pp. 806–809
27. L. Deng, A. Acero, L. Jiang, J. Droppo, X. Huang, High-performance robust speech recognition using stereo training data, in *Proceedings of IEEE International Conference on Acoustics, Speech and Signal Processing*, Salt Lake City, 2001, vol. 1, pp. 301–304
28. L. Buera, E. Lleida, A. Miguel, A. Ortega, Multi-environment models based linear normalization for speech recognition in car conditions, in *Proceedings of IEEE International Conference on Acoustics, Speech and Signal Processing (ICASSP '04)*, Montreal, 2004
29. M. Afify, X. Cui, Y. Gao, Stereo-based stochastic mapping for robust speech recognition. IEEE Trans. Audio Speech Lang. Process. **17**(7), 1325–1334 (2009)
30. L. Mary, B. Yegnanarayana, Extraction and representation of prosodic features for language and speaker recognition. Speech Commun. **50**, 782–796 (2008)
31. A.G. Adami, R. Mihaescu, D.A. Reynolds, J.J. Godfrey, Modeling prosodic dynamics for speaker recognition, in *Proceedings of IEEE International Conference on Acoustics, Speech and Signal Processing (ICASSP '03)*, Hong Kong, 2003
32. L. Ferrer, E. Shriberg, S. Kajarekar, K. Sonmez, Parameterization of prosodic feature distributions for SVM modeling in speaker recognition, in *Proceedings of IEEE International Conference on Acoustics, Speech and Signal Processing (ICASSP '07)*, Honolulu, 2007, pp. 233–236
33. S.G. Koolkagudi, K.S. Rao, R. Reddy, A.K. Vuppala, S. Chakrabarti, Robust speaker recognition in noisy environments: using dynamics of speaker-specific prosody, in *Forensic Speaker Recognition* (Springer, New York, USA, 2013), pp. 183–204
34. E. Shriberg, L. Ferrer, S. Kajarekar, A. Venkataraman, A. Stolckea, Modeling prosodic feature sequences for speaker recognition. Speech Commun. **46**, 455–472 (2005)
35. G. Doddington, Speaker recognition based on idiolectal differences between speakers, in *Proceedings of the European Conference of Speech Communication Technology (EUROSPEECH '01)*, Aalborg, 2001, pp. 2521–2524
36. W.M. Campbel, J.P. Campbell, D.A. Reynolds, D.A. Jones, T.R. Leek, Phonetic speaker recognition with support vector machines, in *Proceedings of the Neural Information Processing Systems Conference*, Vancouver, 2003, pp. 1377–1384
37. K. yee Leung, M. wai Mak, M. Siu, S. yuan Kung, Adaptive articulatory feature-based conditional pronunciation modeling for speaker verification. Speech Commun. **48**, 71–84 (2006)
38. B. Ma, D. Zhu, H. Li, R. Tong, Speaker cluster based GMM tokenization for speaker recognition, in *Proceeding of the 7th Annual Conference of the International Speech Communication Association (INTERSPEECH '06)*, Pittsburgh, 2006
39. B. Ma, H. Li, R. Tong, Spoken language recognition using ensemble classifiers. IEEE Trans. Audio Speech Lang. Process. **15**(7), 2053–2062 (2007)

40. D. Reynolds, W. Andrews, J. Campbell, J. Navratil, B. Peskin, A. Adomi, Q. Jin, D. Kluracek, J. Abramson, R. Mihaescu, J. Godfrey, D. Jones, S. Xiang', The supersid project: exploiting high-level information for high-accuracy speaker recognition, in *Proceedings of IEEE International Conference on Acoustics, Speech and Signal Processing (ICASSP '03)*, Hong Kong, 2003

41. H. Hermansky, Perceptual linear prediction (PLP) analysis for speech. J. Acoust. Soc. Am. **87**, 1738–1752 (1990)

42. L. Rabiner, B.H. Juang, *Fundamentals of Speech Recognition*, 1st edn. (Prentice-Hall, Englewood Cliffs, 1993)

43. X. Huang, A. Acero, H. Hon, *Spoken Language Processing: a Guide to Theory, Algorithm, and System Development* (Prentice Hall, Upper Saddle River, 2001)

44. S. Sarkar, K.S. Rao, D. Nandi, Multilingual speaker recognition on Indian languages, in *IEEE INDICON*, Mumbai (IIT Mumbai, Mumbai, 2013)

45. J.W. Suh, S.O. Sadjadi, G. Liu, T. Hasan, K.W. Godin, J.H. Hansen, Exploring Hilbert envelope based acoustic features in i-vector speaker verification using HT-PLDA, in *Proceedings of NIST Speaker Recognition Evaluation Workshop*, Gaithersburg, USA, 2011

46. C. Kim, R.M. Stern, Power-normalized cepstral coefficients (PNCC) for robust speech recognition, in *Proceedings of IEEE International Conference on Acoustics, Speech and Signal Processing (ICASSP '12)*, Kyoto, 2012

47. V. Mitra, H. Franco, M. Graciarena, A. Mandal, Normalized amplitude modulation features for large vocabulary noise-robust speech recognition, in *Proceedings of IEEE International Conference on Acoustics, Speech and Signal Processing (ICASSP '12)*, Kyoto, 2012

48. A.K. Vuppala, K.S. Rao, Speaker identification under background noise using features extracted from steady vowel regions. Int. J. Adapt. Control Signal Process. **27**(9), 781–792 (2013). Wiley

49. A.K. Vuppala, K.S. Rao, S. Chakrabarti, Improved speaker identification in wireless environment. Int. J. Signal Imaging Syst. Eng. **6**(3), 130–137 (2013)

50. K.S. Rao, S. Maity, V.R. Reddy, Pitch synchronous and glottal closure based speech analysis for language recognition. Int. J. Speech Technol. **16**, 413–430 (2013). Springer

51. T. Kristjansson, B. Frey, Accounting for uncertainity in observations: a new paradigm for robust speech recognition, in *Proceedings of IEEE International Conference on Acoustics, Speech and Signal Processing (ICASSP '02)*, Orlando, 2002, vol. 1, pp. 61–64

52. C.H. Lee, On stochastic feature and model compensation approaches to robust speech recognition. Speech Commun. **25**, 29–47 (1998)

53. C.H. Lee, Q. Huo, On adaptive decision rules and decision parameter adaptation for automatic speech recognition. Proc. IEEE **88**(8), 1241–1269 (2000)

54. T. Quatieri, D. Reynolds, G. O'Leary, Estimation of handset nonlinearity with application to speaker recognition. IEEE Trans. Speech Audio Process. **8**, 567–584 (2000)

55. H.A. Murthy, F. Beaufays, L.P. Heck, M. Weintraub, Robust text-independent speaker identification over telephone channels. IEEE Trans. Speech Audio Process. **7**(5), 554–568 (1999)

56. R. Teunen, B. Shahshahani, L. Heck, A model-based transformational approach to robust speaker recognition, in *Proceeding of the Annual Conference of the International Speech Communication Association (INTERSPEECH '00)*, Beijing, 2000, vol. 2, pp. 495–498

57. J. Gauvain, C. Lee, Maximum a posteriori estimation for multivariate Gaussian mixture observations of Markov chains. IEEE Trans. Speech Audio Process. **2**(2), 291–298 (1994)

58. C. Leggetter, P. Woodland, Maximum likelihood linear regression for speaker adaptation of continuous density HMMs. Comput. Speech Lang. **9**, 171–185 (1995)

59. D.A. Reynolds, R.C. Rose, Robust text-independent speaker identification using Gaussian mixture speaker models. IEEE Trans. Acoust. Speech Signal Process. **3**(1), 72–83 (1995)

60. D. Reynolds, T. Quatieri, R. Dunn, Speaker verification using adapted Gaussian mixture models. Digit. Signal Process. **10**(1), 19–41 (2000)

61. D. Zhu, B. Ma, H. Li, Joint MAP adaptation of feature transformation and Gaussian mixture model for speaker recognition, in *Proceedings of IEEE International Conference on Acoustics, Speech and Signal Processing (ICASSP '09)*, Taipei, 2009, pp. 4045–4048

62. C.M. Bishop, *Pattern Recognition and Machine Learning* (Springer, New York, 2006)
63. V. Digalakis, D. Rtischev, L. Neumeyer, E. Sa, Speaker adaptation using constrained estimation of Gaussian mixtures. IEEE Trans. Speech Audio Process. **3**(5), 357–366 (1995)
64. S. Kozat, K. Visweswariah, R. Gopinath, Feature adaptation based on Gaussian posteriors, in *Proceedings of IEEE International Conference on Acoustics, Speech and Signal Processing*, Toulouse, 2006, pp. 221–224
65. K.K. Yiu, M.W. Mak, S.Y. Kung, Environment adaptation for robust speaker verification, in *Proceedings of the European Conference of Speech Communication and Technology (EUROSPEECH '03)*, Geneva, 2003, vol. 2, pp. 2973–2976
66. M.J.F. Gales, S.J. Young, Robust speech recognition in additive and convolutional noise using parallel model combination. Comput. Speech Lang. **9**, 289–307 (1995)
67. L.P. Wong, M. Russell, Text-dependent speaker verification under noisy conditions using parallel model combination, in *Proceedings of IEEE International Conference on Acoustics, Speech and Signal Processing (ICASSP '01)*, Salt Lake City, 2001, pp. 457–460
68. P. Moreno, Speech recognition in noisy environments. PhD thesis, Electrical & Computer Engineering Department, Carnegie Mellon University, Pittsburgh, 1996
69. K.C. Sim, M.T. Luong, A trajectory-based parallel model combination with a unified static and dynamic parameter compensation for noisy speech recognition, in *Proceedings of the Workshop on Automatic Speech Recognition and Understanding (ASRU '11)*, Waikoloa, Dec 2011, pp. 107–112
70. P.J. Moreno, B. Raj, R.M. Stern, A vector Taylor series approach for environment-independent speech recognition, in *Proceedings of IEEE International Conference on Acoustics, Speech and Signal Processing*, Atlanta, 1996, pp. 733–736
71. A. Sankar, C.H. Lee, Stochastic matching for robust speech recognition. IEEE Signal Process. Lett. **1**(8), 124–125 (1994)
72. H. Liao, M.J.F. Gales, Joint uncertainty decoding for noise robust speech recognition, in *Proceedings of 6th Annual Conference of the International Speech Communication Association (INTERSPEECH '05)*, Lisbon, 2005
73. J. Ming, D. Stewart, S. Vaseghi, Speaker identification in unknown noisy conditions – a universal compensation approach, in *Proceedings of IEEE International Conference on Acoustics, Speech and Signal Processing (ICASSP '05)*, Philadelphia, 2005
74. J. Ming, T.J. Hazen, J.R. Glass, D. Reynolds, Robust speaker recognition in noisy conditions. IEEE Trans. Audio Speech Lang. Process. **15**(5), 1711–1723 (2007)
75. A. Drygajlo, M. El-Maliki, Speaker verification in noisy environment with combined spectral subtraction and missing data theory, in *Proceedings of IEEE International Conference on Acoustics, Speech and Signal Processing (ICASSP '98)*, Seattle, 1998
76. D. Burton, Text-dependent speaker verification using vector quantization source coding. IEEE Trans. Acoust. Speech Signal Process. **35**(2), 133–143 (1987)
77. T. Kinnunen, E. Karpov, P. Franti, Real-time speaker identification and verification. IEEE Trans. Audio Speech Lang. Process. **14**(1), 277–288 (2006)
78. M.W. Mak, R. Hsiao, B. Mak, A comparison of various adaptation methods for speaker verification with limited enrollment data, in *Proceedings of IEEE International Conference on Acoustics, Speech and Signal Processing (ICASSP '06)*, Toulouse, 2006, pp. 929–932
79. V. Hautamaki, T. Kinnunen, I. Karkkainen, M. Tuononen, J. Saastamoinen, P. Franti, Maximum a posteriori adaptation of the centroid model for speaker verification. IEEE Signal Process. Lett. **15**, 162–165 (2008)
80. P. Kenny, G. Boulianne, P. Ouellet, P. Dumouchel, Factor analysis simplified, in *Proceedings of the IEEE International Conference on Acoustics Speech and Signal Processing (ICASSP '05)*, Philadelphia, 2005, vol. 1, pp. 637–640
81. K. Farrell, R. Mammone, K. Assaleh, Speaker recognition using neural networks and conventional classifiers. IEEE Trans. Speech Audio Process. **2**(1), 195–204 (1994)
82. L.P. Heck, Y. Konig, M. Sonmez, M. Weintraub, Robustness to telephone handset distortion in speaker recognition by discriminative feature design. Speech Commun. **31**, 181–192 (2000)

83. B. Yegnanarayana, S.P. Kishore, AANN: an alternative to GMM for pattern recognition. Neural Netw. **15**, 456–469 (2002)
84. W. Campbell, J. Campbell, D. Reynolds, E. Singer, P. Carrasquillo, Support vector machines for speaker and language recognition. Comput. Speech Lang. **20**, 210–229 (2006)
85. W. Campbell, J. Campbell, D. Reynolds, Support vector machines using GMM supervectors for speaker verification. IEEE Signal Process. Lett. **13**(5), 308–311 (2006)
86. V. Wan, S. Renals, Speaker verification using sequence discriminant support vector machines. IEEE Trans. Acoust. Speech Audio Process. **13**(2), 203–210 (2005)
87. C.H. You, K.A. Lee, H. Li, An SVM kernel with GMM-supervector based on the Bhattacharyya distance for speaker recognition. IEEE Signal Process. Lett. **16**(1), 49–52 (2009)
88. A. Solomonoff, C. Quillen, I. Boardman, Channel compensation for SVM speaker recognition, in *IEEE Workshop on Speaker and Language Recognition (Odyssey '04)*, Toledo, 2004, pp. 57–62
89. A.O. Hatch, S. Kajarekar, A. Stolcke, Within-class covariance normalization for SVM-based speaker recognition, in *Proceedings of the International Conference of Spoken Language Processing (ICSLP '05)*, Lisbon, Portugal, 2005
90. P. Kenny, G. Boulianne, P. Dumouchel, Eigenvoice modeling with sparse training data. IEEE Trans. Speech Audio Process. **13**(3), 345–354 (2005)
91. N. Dehak, R. Dehak, P. Kenny, N. Brummer, P. Ouellet, P. Dumouchel, Support vector machines versus fast scoring in the low-dimensional total variability space for speaker verification, in *Proceeding of the 10th Annual Conference of the International Speech Communication Association (INTERSPEECH '09)*, Brighton, 2009
92. N. Dehak, P. Kenny, R. Dehak, O. Glembek, P. Dumouchel, L. Burget, V. Hubeika, F. Castaldo, Support vector machines and joint factor analysis for speaker verification, in *Proceedings of IEEE International Conference on Acoustics, Speech and Signal Processing (ICASSP '09)*, Taipei, 2009, pp. 4237–4240
93. M. McLaren, D. van Leeuwen, Source-normalized LDA for robust speaker recognition using i-vectors from multiple speech sources. IEEE Trans. Audio Speech Lang. Process. **20**(3), 755–766 (2012)
94. T. Kinnunen, Spectral features for automatic text-independent speaker recognition. PhD thesis, Department of Computer Science, University of Joensuu, 2004
95. N. Dehak, P.J. Kenny, R. Dehak, P. Dumouchel, P. Ouellet, Front-end factor analysis for speaker verification. IEEE Trans. Audio Speech Lang. Process. **19**(4), 788–798 (2011)

# Chapter 3
# Speaker Verification in Noisy Environments Using Gaussian Mixture Models

**Abstract** This chapter explores the behavior of Gaussian Mixture Models (GMMs) for speaker verification in noisy environments. Specifically, the performance of an acoustic modeling framework (namely GMM-UBM) using speaker-dependent GMMs and a speaker-independent Universal Background Model (UBM), is studied for simulated noisy backgrounds. Significance of a feature mapping technique using multiple UBMs for compensating background noise is explored. The speaker verification systems explored in this chapter serve the purpose of baselines considered for comparison and analyzing the performance improvements of the proposed methods in the remaining chapters.

## 3.1 Introduction

The role of acoustic modeling for speaker recognition (SR) was briefly introduced in Chaps. 1 and 2, respectively. Specifically, one or more types of statistical models are employed to capture the unique distribution of features extracted from a speaker's enrollment utterances during the training phase. During the recognition phase, an unknown utterance is classified as a speaker based on its similarities with the corresponding speaker model. Effectiveness of a model is characterized by its classification accuracy, computational costs, data requirements etc. In most cases, selection of a suitable model is a tradeoff between one or more such criteria.

Gaussian Mixture Models (GMMs) are the most extensively used speaker modeling techniques in text-independent speaker verification (SV) [1]. They belong to the family of generative models in which a number of non-uniformly weighted multivariate Gaussian components are used to represent the feature distribution of an individual speaker. GMMs have been found to effectively characterize multi-modal spectral shapes and model arbitrary spectral densities. Besides providing a strong probabilistic framework for pattern matching, they offer large degrees of freedom and have high recognition accuracy [2].

K.S. Rao and S. Sarkar, *Robust Speaker Recognition in Noisy Environments*,
SpringerBriefs in Electrical and Computer Engineering,
DOI 10.1007/978-3-319-07130-5_3, © The Author(s) 2014

This chapter explores the behavior of Gaussian Mixture Models (GMMs) for speaker verification in noisy environments. Specifically, the performance of an acoustic modeling framework (namely GMM-UBM) using speaker-dependent GMMs and a speaker-independent Universal Background Model (UBM), is studied for simulated noisy backgrounds. Significance of a feature mapping technique using multiple UBMs for compensating background noise is explored. The SV systems explored in this chapter serve the purpose of baselines considered for comparison and analyzing the performance improvements of the proposed methods in the remaining chapters. The rest of the chapter is organized as follows. Section 3.2 describes the GMM-UBM framework, the SV system development is discussed in Sect. 3.3, the feature mapping method in Sect. 3.4.2 followed by a brief summary in Sect. 3.6.

## 3.2 GMM-UBM Framework for Speaker Verification

The Speaker Verification (SV) process was briefly introduced in Chap. 1. It is the task of validating the claimed identity of a person using his/her speech. In the conventional SV paradigm the task can be viewed as a binary classification problem in which a claimant's utterance is classified as true (authentic) or false (impostor) based on its statistical similarities with a claimed (target) speaker model and an impostor model, respectively. The target speaker model essentially belongs to one out of a group of speakers enrolled for the SV system while the common impostor model is constructed offline from selected set of impostors.

Operation of the GMM-UBM framework [3] is similar to the standard SV process which includes an offline phase for impostor/background model construction and an online phase comprising enrollment and verification. A single GMM (namely UBM), is constructed during the offline phase using substantial amount of data collected from various speakers across multiple channels, environments etc. GMMs are particularly well suited for this task since the model complexity can be scaled to handle large datasets. This is simply achieved by using a large number of mixture components (typically 1,024, 2,048 etc.). The Gaussian components are considered to model various acoustic events e.g., broad phonetic sounds that characterize a person's voice. The data used for model construction is specifically chosen to encompass various acoustic conditions so that the UBM can be used for generalized SV tasks. For a $D$-dimensional feature vector ($y$) from the training set, the UBM ($\lambda$) likelihood function is defined as a weighted sum of $M$ component Gaussian densities as given by

$$p(y|\lambda) = \sum_{i=1}^{M} w_i \, p_i(y) \tag{3.1}$$

where $w_i$ are the mixture weights and $p_i(y)$ are the component densities. Each component density in turn is a $D$-variate Gaussian function of the form

$$p_i(y) = \frac{1}{(2\pi)^{D/2}\,|\Sigma_i|^{\frac{1}{2}}}\, \exp\left\{-\frac{1}{2}(y-\mu_i)^T\,\Sigma_i^{-1}\,(y-\mu_i)\right\} \tag{3.2}$$

where $\mu_i$ and $\Sigma_i$ denote the mean vector and covariance matrix for the $i$th component. The mixture weights are furthermore required to satisfy the constraint $\sum_{i=1}^{M} w_i = 1$. The complete GMM, parameterized by the mean vectors, covariance matrices and the mixture weight from all component densities is collectively represented by $\lambda = \{w_i, \mu_i, \Sigma_i\}$; where $i = 1, 2, \ldots M$. The GMM parameters are estimated by iteratively maximizing the likelihood of the training data using an Expectation Maximization (EM) algorithm [4]. Details of the GMM training procedure has been outlined in Appendix B.

As the name implies, the UBM represents a speaker-independent impostor model. The SV process essentially requires construction of target speaker models (GMMs) for representing the true (actual) speaker class. However, maximum likelihood GMM training for individual target speakers would require an adequate amount of training data from each of them which is considered unfavourable for practical scenarios (e.g., real-time SV systems). As a feasible alternative, target speaker models (GMMs) are constructed by a *Maximum aPosteriori* (MAP) adaptation [5] of the existing UBM parameters, during the enrollment phase. The MAP adaptation procedure consists of two broad stages i.e., estimation of a set of sufficient statistics from a target speaker's enrollment data followed by actual modification (adaptation) of the UBM parameters. The overall adaptation procedure can be briefly outlined in the following steps. Given a sequence of **T** training vectors $(x_1, x_2, \ldots, x_T)$ extracted from a target speaker utterance, the posterior probability of each mixture $i$ in the UBM with respect to a vector $x_t$, is calculated as

$$\Pr(i|x_t) = \frac{w_i\, p_i(x_t)}{\sum_{j=1}^{M} w_j\, p_j(x_t)} \tag{3.3}$$

Using $Pr(i|x_t)$, a set of sufficient statistics are calculated using the training set as follows

$$n_i = \sum_{t=1}^{T} \Pr(i|x_t)$$

$$E_i(x) = \frac{\sum_{t=1}^{T} \Pr(i|x_t)x_t}{n_i}$$

$$E_i(x^2) = \frac{\sum_{t=1}^{T} \Pr(i|x_t)x_t^2}{n_i} \tag{3.4}$$

These new statistics are used to alter the pre-estimated UBM parameters. A new set of model parameters i.e., weights ($\hat{w}_i$), means ($\hat{\mu}_i$) and variances ($\hat{\sigma}_i^2$), for the target speaker model ($\lambda_{Tar}$) are obtained as

$$\hat{w}_i = \left[ \frac{\alpha_i n_i}{T} + (1 - \alpha_i) w_i \right] \gamma$$
$$\hat{\mu}_i = \alpha_i E_i(x) + (1 - \alpha_i) \mu_i$$
$$\hat{\sigma}_i^2 = \alpha_i E_i(x^2) + (1 - \alpha_i)(\sigma_i^2 + \mu_i^2) - \hat{\mu}_i^2 \qquad (3.5)$$

where $\gamma$ is a scaling factor, which ensures that all the new mixture weights sum to 1 and $\alpha_i$ is an adaptation coefficient which controls the balance between the old and new model parameter estimates, defined as

$$\alpha_i = \frac{n_i}{n_i + r} \qquad (3.6)$$

where $r$ is a fixed relevance factor, which determines the extent of mixing of the old and new estimates of the parameters. Low values for $\alpha_i$ ($\alpha_i \to 0$), results in negligible modification of the UBM parameters, while higher values ($\alpha_i \to 1$) strictly emphasize on the use of new training data-dependent parameters. Besides minimal data requirements, the MAP adaptation has other interesting aspects. Firstly, derivation of speaker model parameters from the well-trained UBM parameters provides a tight coupling between the UBM and the speaker model. Besides providing better performance than de-coupled models, it allows scope for a fast scoring technique during the evaluation stage [3]. Secondly, it offers extra robustness to the SV system by adapting model parameters to reflect new acoustic conditions during the enrollment phase. Figure 3.1 illustrates the GMM-UBM framework for speaker verification. Given the UBM ($\lambda$) and a target speaker model ($\lambda_{Tar}$), SV reduces to a hypothesis testing problem in which the task is to decide whether an unknown test utterance is generated from the hypothesized target speaker model (null hypothesis) or the UBM (alternative hypothesis). During the evaluation phase, a log-likelihood ratio of scores ($S(X_{test})$) generated by the two models from a given set of test feature vectors ($X_{test}$) (as shown in Eq. 3.7), is compared with a empirically determined threshold for acceptance or rejection.

$$S(X_{test}) = \log p(X_{test}|\lambda_{Tar}) - \log p(X_{test}|\lambda) \qquad (3.7)$$

where $X_{test} = \{x_1, x_2, \ldots, x_K\}$ and $p(X_{test}|\Lambda) = \sum_{k=1}^{K} log(p(x_k|\Lambda)); \Lambda = \{\lambda, \lambda_{Tar}\}$.

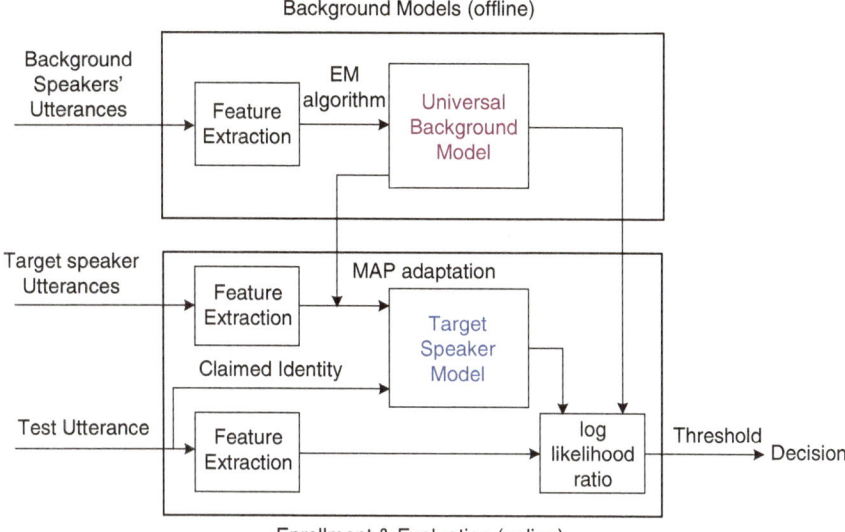

**Fig. 3.1** The GMM-UBM framework for speaker verification

## 3.3   Development of GMM-UBM Based Speaker Verification Systems

The NIST-2003-SRE database [6] was used for developing all SV systems. The data consists of conversational speech collected from 356 speakers (149 Male and 207 Female) over a cellular phone network. Each speech file is sampled at 8 kHz with a bit resolution of 16 bits/sample. The training set contains approximately 2 min of one-sided conversational speech from each enrolled speaker. The test set contains around 3,500 speech segments of approximately 15 s each.

### 3.3.1   Simulation of Background

As mentioned earlier in Chap. 1, the goal of the current work is to explore the impact of environmental noise on the performance of the SV systems. However, for experimental convenience and due to unavailability of real-life noisy data, the noisy backgrounds were artificially simulated. All training and test utterance were degraded with additive noises collected from the NOISEX-92 database [7]. Two types of backgrounds (i.e., uniform and time-varying) were independently considered for SV studies. Each of these are described in the following subsections

#### 3.3.1.1 Simulation of Uniform Background

Four additive noises (i.e., car, factory, pink and white) collected from the NOISEX-92 database was used for representing unique background environments. The speech segment from each of the 356 enrolled speakers was degraded by adding a specific type of noise at 0, 5 and 10 dB SNRs, respectively. The noise level was scaled to maintain the desired SNRs of the reconstructed speech segments. Twelve different sets of noisy training utterances were obtained (one for each noise at a particular SNR). Each set was used separately for training and evaluation.

#### 3.3.1.2 Simulation of Varying Background

To capture the effect of time varying environmental noise, each enrollment utterance was firstly divided into five non-overlapping segments. Each segment was individually corrupted with one of four additive noise chosen randomly. The remaining one segment was left clean. The reconstruction process was repeated for 0, 5, 7 and 10 dB SNRs, respectively. For a relative reduction of mismatch all the test utterances were similarly corrupted by the same noises. Four different training and test sets of time-varying noisy utterances were obtained (one for each SNR). Each set corresponding to a particular SNR was used separately for training and evaluation. Additionally, another set of clean training utterances was maintained for experimenting with a mismatched condition as discussed in Sect. 3.3.5.

### 3.3.2 Preprocessing

The speech signals were pre-emphasized using a first order high pass filter with a coefficient of 0.97. The resultant signals were processed using short-term frames of 20 ms with a frame-overlap of 10 ms. A simple energy-based thresholding scheme was used for voiced activity detection (VAD) as described in [8]. For the simulated stereo data used in the present work, the VAD was applied in two stages. The average energy of an entire clean speech utterance ($E_{avg}$) was calculated and a threshold of 1 % of the average energy ($0.01 \times E_{avg}$) was empirically determined. All clean frames with an average energy below the selected threshold were discarded. The frame indices of the discarded frames were recorded for later use. After noise contamination of the clean utterance as described in Sect. 3.3.1, the noisy frames corresponding to recorded indices were discarded without any further thresholding.

### 3.3.3 Feature Extraction

Mel frequency cepstral coefficient (MFCC) features were used throughout the development process. A 26 channel mel-scaled triangular filterbank constrained in the telephonic bandwidth of 300–3,400 Hz, was imposed on the DFT magnitude spectra of each frame. Thirteen cepstral coefficients ($C_1 - C_{13}$) excluding the zeroth one ($C_0$), were extracted after discrete cosine transform of the log filterbank energies. The delta and acceleration coefficients computed over a frame span of 2, were appended to form a 39-dimensional feature vector. All feature vectors were subjected to cepstral mean subtraction followed by cepstral variance normalization. The resultant distribution was scaled to zero mean and unit variance. A distinct advantage of using MFCCs apart from their compact and discriminative properties, is that individual features are highly de-correlated which allows the usage of diagonal covariance during GMM modeling minimizing data requirements and computational load. The MFCC extraction process is explained in more details in Appendix C. In the ensuing discussions, the terms 'features' and 'data' have been used interchangeably.

### 3.3.4 Speaker Modeling

Acoustic modeling using the standard GMM-UBM framework was performed in two stages i.e., construction of a Universal Background Model (UBM) and the target speaker models. The SwitchBoard corpus (part II) was used for UBM construction. The data consisting of 20 h of conversational speech from 100 male and 100 female speakers (6 min from each speaker) was subjected to preprocessing and feature extraction. The MFCC vectors were pooled together into a single dataset. Vector quantization (VQ) was initially used to divide the dataset into 1,024 clusters. The VQ centroids and fraction of data occupying each cluster were used to initialize the GMM means and weights, respectively. The covariances were initialized to identity matrices. A gender-independent 1,024-component GMM (UBM) was trained offline on the dataset using 200 iterations of the EM algorithm.

The target speaker models (GMMs) were derived individually, by MAP adaptation of the UBM using the noisy enrollment utterances from each target speaker in each dataset, according to Eqs. (3.4) and (3.7), respectively. The relevance factor for calculating the parameter $\alpha$ was fixed at $r = 16$. Thus, 356 target speaker GMMs were obtained from each noisy dataset constructed in Sect. 3.3.1. Model parameters adapted to residual channel noises present in the original NIST training data expectedly offered extra robustness to the SV systems.

### 3.3.5 Performance Evaluation

In this NIST-2003 primary task, each test segment (unknown claimant) is evaluated against 11 hypothesized target speaker models (GMMs) each of which represents the claimed identity of the test speaker. In each set of 11 trials (evaluation), at most one is true (actual speaker) while the rest are false (impostor). Ideally, the log-likelihood ratio score obtained in each trial is supposed to be compared with a pre-determined threshold for final acceptance (true score) or rejection (false score), as discussed in Sect. 3.2. A SV system is then susceptible to two types of error i.e., a 'false alarm' when an impostor's claim is accepted and 'miss' when a true identity is rejected.

Alternatively, the NIST-SRE-2003 evaluation requires calculation of two metrics, namely Equal Error Rate (EER) and Detection Cost Function (DCF). The probabilities of the two aforementioned errors are calculated from the true and false scores, respectively. A Detection Error Tradeoff (DET) curve [9] is obtained by plotting the 'miss' and 'false alarm' probabilities for all trials in a normal-deviate scale. The Equal Error Rate (EER), defined as the operating point in the DET curve where both the miss and false alarm rates are equal, is recorded. The DCF is calculated as a weighted sum of the two error probabilities as follows

$$DCF = C_{Miss} P_{Miss} P_{Target} + C_{FalseAlarm} P_{FalseAlarm}(1 - P_{Target})$$

where cost of miss $C_{Miss} = 10$, cost of false alarm $C_{FalseAlarm} = 1$, probability of target $P_{Target} = 0.01$, $P_{Miss} =$ probability of miss and $P_{FalseAlarm} =$ probability of false alarm. The EER and minimum DCF (MinDCF) values were used as metrics for performance evaluation.

Primary tasks were individually carried out on the original NIST data and each simulated noisy dataset as described in Sect. 3.3.1 in clean, matched and mismatched conditions, respectively. Each of these conditions are defined as follows

- Clean: Clean condition refers to the ideal scenario when both enrollment (training) and verification (testing) phases are carried out in environments free of noise distortions. The primary task carried out on the original NIST data defines this condition.
- Matched: This refers to the default condition where both training and testing phases of SV are carried out in similar noisy environments. The test utterances in each noisy dataset were evaluated against the target speaker models built using similar type of noisy data.
- Mismatched: The mismatch occurs due to clean training and noisy testing environments. Clean speaker models were constructed using the original NIST enrollment utterances. The noisy test utterances in each dataset were evaluated against these models in individual experiments according to the primary task. This condition was studied for both uniform and varying background environments.

Performance of the SV system developed in clean condition is given in Table 3.1. EER and MinDCF values at 6.93 % and 0.033, respectively were obtained by the default configuration i.e., primary task on NIST data. These results are later used for comparing and contrasting performances of SV systems in matched and mismatched conditions.

Table 3.2 summarizes the performance of the SV systems in uniform background environments in matched and mismatched conditions, respectively. An overall

Table 3.1 Performance of the GMM-UBM based SV system in clean conditions

| Condition | EER (%) | MinDCF |
|-----------|---------|--------|
| Clean     | 06.93   | 0.033  |

Table 3.2 Summary of performance of the GMM-UBM based SV systems in uniform background environments

| Condition | Noises | SNR (0 dB) | | SNR (5 dB) | | SNR (10 dB) | |
|-----------|--------|---------|--------|---------|--------|---------|--------|
|           |        | EER (%) | MinDCF | EER (%) | MinDCF | EER (%) | MinDCF |
| Matched   | Car     | 18.04 | 0.071 | 18.11 | 0.071 | 15.44 | 0.068 |
|           | Factory | 23.17 | 0.089 | 20.96 | 0.083 | 16.44 | 0.072 |
|           | Pink    | 26.65 | 0.092 | 23.89 | 0.092 | 18.65 | 0.081 |
|           | White   | 30.98 | 0.097 | 27.41 | 0.094 | 21.91 | 0.087 |
| Mismatched | Car    | 20.55 | 0.079 | 20.95 | 0.077 | 10.75 | 0.049 |
|           | Factory | 32.16 | 0.099 | 27.15 | 0.098 | 16.53 | 0.076 |
|           | Pink    | 35.05 | 0.099 | 30.39 | 0.097 | 18.83 | 0.085 |
|           | White   | 39.02 | 0.099 | 34.16 | 0.099 | 21.77 | 0.092 |

observation reveals drastic increments in both the error metrics across all environments, thereby implying the sensitivity of the GMM-UBM framework towards noise distortions. The performances in distinct backgrounds are characterized by the type of noise present in each of them. Two typical aspects of the SV performance are notable irrespective of the noisy backgrounds. Firstly, a consistent degradation of performance accuracy is observed across all backgrounds with decreasing SNRs with an exception in case of car noise at 0 and 5 dB. Secondly, it is interesting to note that the performances in matched conditions are comparatively much better than those in mismatched conditions. Specifically, average EER increments of 0.66, 15.27, 15.08 and 14.65 % are observed in the mismatched cases for car, factory, pink and white noisy backgrounds across all SNRs, respectively. An anomalous behavior is once again seen in case of mismatched conditions for car noise at 10 dB. A general order of precedence of the various backgrounds (i.e., car, factory, pink and white) is apparent in terms of improving performances. This is evident from the average EER values across all SNRs at 17.20, 20.19, 23.06, 26.77 % and 17.42, 25.28, 28.09, 31.65 % for car, factory, pink and white noisy backgrounds in matched and mismatched conditions, respectively.

Figure 3.2 demonstrates the DET plots for the SV systems developed in uniform background environments. Each set of curves in a subfigure corresponds to a SV evaluation in a particular type of background noise for matched and mismatched conditions across various SNRs. Some general characteristics are observable in each subfigure. The DET curves consistently shift towards the origin with increasing SNRs implying performance improvements with decreasing noise strength. As discussed earlier, the mismatched conditions (represented by broken lines) show inferior performance to the matched conditions in a particular SNR apart from the

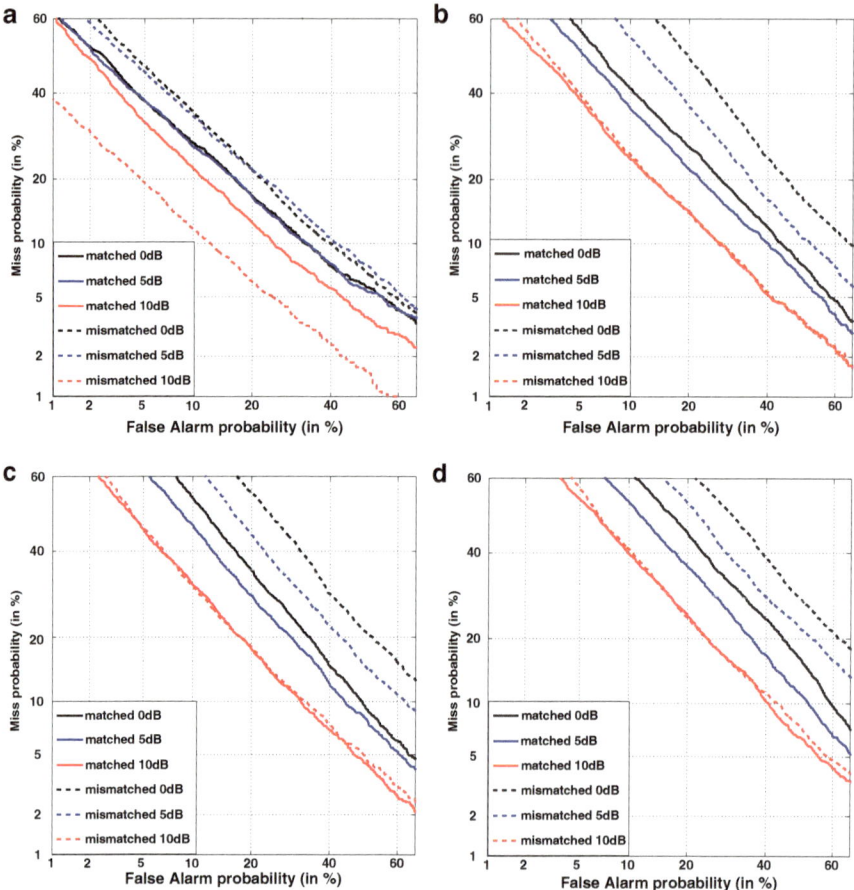

**Fig. 3.2** DET plots of the SV systems developed in uniform background environment with (**a**) car (**b**) factory (**c**) pink and (**d**) white noise. The *black, blue* and *red colored curves* in each plot indicate SNR levels of 0, 5, and 10 dB, respectively. *Solid* and *broken lines* indicate matched and mismatched conditions, respectively

prominent anomaly seen in case of car noise at 10 dB. The effect of mismatch is significantly subdued with SNR increments as seen by the overlap in the red curves. This effect is particularly prominent in case of factory, white and pink noisy backgrounds. Each set of curves (corresponding to a specific noise) show distinct characteristics in terms of slope and alignment of operating points.

Table 3.3 summarizes the performance of the SV systems developed in varying background environments in the three distinct conditions. Unlike the uniform background scenario, the present results reflect the system's behavior in a non-stationary environment. As observed earlier, the SV performances degrade consistently with decreasing SNRs. Similarly, inferior performances are observed in mismatched condition in comparison to matched condition. The effect of the mismatch is most

**Table 3.3** Summary of performance of the GMM-UBM based SV systems in varying background environments

| SNRs | Matched | | Mismatched | |
|---|---|---|---|---|
| (dB) | EER (%) | MinDCF | EER (%) | MinDCF |
| 0 | 27.05 | 0.094 | 32.25 | 0.099 |
| 5 | 25.74 | 0.086 | 30.89 | 0.098 |
| 7 | 25.29 | 0.083 | 28.09 | 0.095 |
| 10 | 21.86 | 0.080 | 25.38 | 0.091 |

prominent in case of 0 dB SNR and decreases gradually with SNR increments. Specifically, the EER increments relative to the matched condition are 5.20, 5.15, 2.80 and 3.52 % for 0, 5, 7 and 10 dB SNRs, respectively.

Figure 3.3 demonstrates the DET plots of the SV systems developed in varying background environments. Each set of curves in a subfigure corresponds to the SV evaluation in a particular SNR. As observed earlier, the curves corresponding to matched condition show inferior performance accuracy in comparison to the mismatched conditions. Unlike the uniform background scenario, a rotation in the curves can be noticed in the mismatched conditions especially at 7 and 10 dB SNRs, respectively. Though a direct comparison is inappropriate, the SV performances in the varying background scenarios are observed to be inferior compared the uniform background at equivalent SNRs. Specifically, average EERs across 0, 5 and 10 dB SNRs show increments of 3.08 and 3.90 % in matched and mismatched conditions, respectively. Thus, a feature compensation technique was further explored for the SV in varying background environments.

## 3.4   Feature Mapping for Speaker Verification in Varying Background Environment

Popular techniques for noise compensation at an acoustic model level such as parallel model combination [10] or Vector Taylor Series [11] assume prior availability of a clean speaker model or a statistical model of the background noise which is often not suitable for practical scenarios. Speaker model synthesis (SMS) [12] provides a more realistic compensation framework by constructing speaker models from multiple unseen channel data by mean shift, variance scaling and weight scaling of a pre-trained model parameters. Feature mapping proposed in [13] uses the SMS framework for mapping individual channel dependent features to a channel independent space by parametric scaling. Being a feature level approach, it has the advantage of being independent of any particular recognition model. As discussed in details in Sect. 3.4.2, the feature mapping technique is used for frame-wise transformation of features extracted from time-varying noisy utterances to achieve an overall denoising impact.

Recent studies demonstrate the merits of using multiple background models (BMs) as an alternative to a single universal background model (UBM) for SV [14].

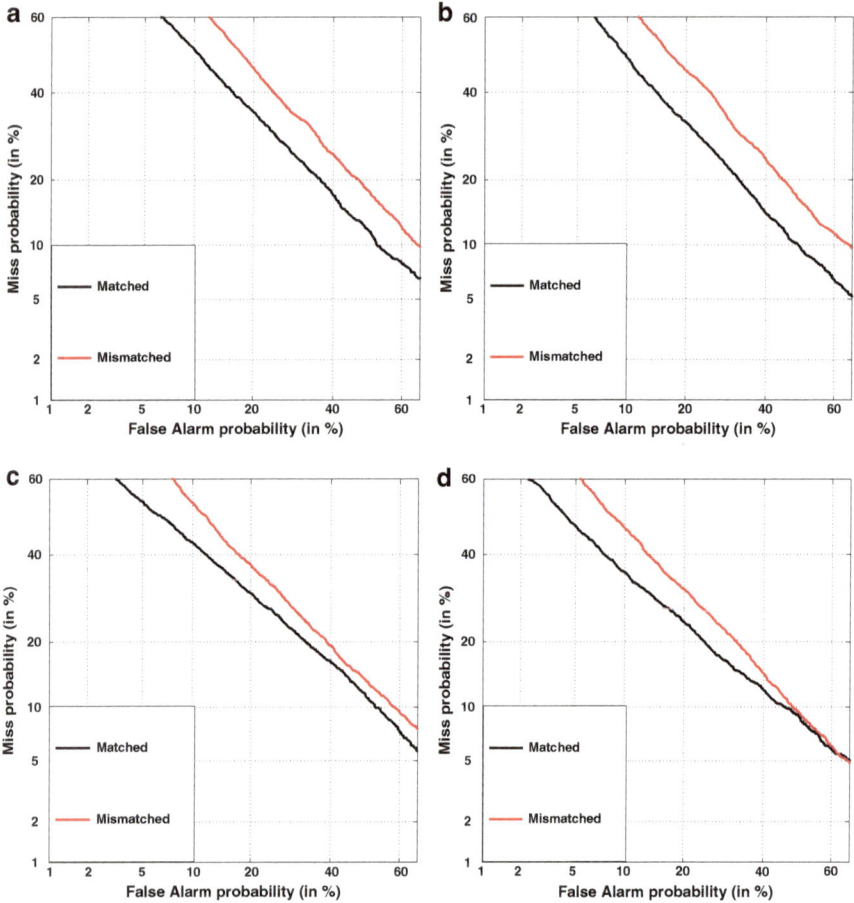

**Fig. 3.3** DET plots of the SV systems developed in varying background environments at (**a**) 0 dB (**b**) 5 dB (**c**) 7 dB and (**d**) 10 dB SNR levels. The *black* and *red colored curves* in each plot indicate matched and mismatched conditions, respectively

The key idea is to construct individual BMs for a group of target speakers with same vocal tract length (VTL) factor for better speaker-impostor discrimination. In a later study [15], the feature mapping technique was integrated with VTL-wise multiple background model framework for exploiting the benefits of both. Instead of independent construction, the background models were derived by MAP adaptation of a root UBM using VTL-wise clustered speaker data. During enrollment and evaluation the most likely BM is selected and the features are mapped to the common/root UBM. The present work uses the method for mapping noisy training and test data to a noise-independent space and demonstrate that the resultant features obtained have a higher statistical match to clean speaker models.

### 3.4.1 Multiple Background Models Based on Vocal Tract Length

The standard GMM-UBM method for speaker verification, as discussed in Sect. 3.2, requires training of a Gaussian Mixture Model (GMM) using a large amount of speech data from various speakers across multiple channels [3]. The so called Universal Background Model (UBM) characterizes speaker independent feature distribution and represents the impostor class for speaker verification (SV). However due to aggregation of data irrespective of gender, age and other speaker-specific information, a single UBM is sometimes inappropriate for generalized SV tasks [14].

The vocal tract length (VTL) is a prominent factor responsible for inter-speaker variability. It was studied in [16] that the differences in VTL between speakers can be resolved by warping the frequency axis of a speaker spectra $S_A(f)$ to that of a reference speaker spectra $S_B(f)$ by a piece-wise linear or non-linear function associated with a VTL warping factor $\alpha$ as shown in Eq. (3.8). Spectral features extracted by filterbank analysis using the warped spectrum is known as vocal tract length normalization (VTLN). This method is used for improving speech recognition tasks where speaker information is suppressed. Depending on a speaker's VTL, the discretized values of $\alpha$ ranging from 0.80 to 1.20 with a step-wise increment of 0.02 in succession are used for deriving the speaker models. Due to unavailability of the reference speaker model, the optimum value of $\alpha$ for a particular speaker is estimated by maximizing the likelihood of the warped features, over the given range of alpha, against the speaker independent UBM as shown in Eq. (3.9).

$$S_A(f^\alpha) = \begin{cases} \alpha f & 0 \le f \le f_0 \\ \frac{f_{max}-\alpha f_0}{f_{max}-f_0}(f-f_0)+\alpha f_0 & f_0 < f \le f_{max} \end{cases} \qquad (3.8)$$

where $S_B(f) = S_A(f^\alpha)$ is the spectra $S_A(f)$ warped by $\alpha$, $f_0$ and $f_{max}$ are the minimum and maximum signal bandwidth.

$$\alpha^* = \arg \max_{\alpha} p(X_s^\alpha | \Lambda_{UBM}) \qquad (3.9)$$

where $X_s^\alpha$ are the features extracted from the spectra scaled by $\alpha$ and $\Lambda_{UBM}$ is the root UBM. Since differences in VTL factor is itself a distinguishing property for speakers, it was used directly for speaker recognition in [17]. A different approach was followed in [14] where the entire training corpus was partitioned in datasets according to the common VTL factor ($\alpha$) of speakers in each dataset. A VTL dependent background model (VTL-BM) was constructed from each dataset. Instead of a single UBM, multiple VTL-BMs were used for the SV task. It was shown that the multiple background models were more effective in rejecting close impostors for an overall improvement in performance accuracy of SV. However as mentioned in [15], a major drawback of using VTL-BMs is the difficulty in applying score normalization techniques like AT-norm and Z-norm. It requires maintenance of individual sets of score normalization speakers for each dataset. To address this drawback and to implement feature mapping using VTL-BMs, a different approach was proposed in [15], where the VTL-BMs were derived by MAP adaptation of a single root UBM using the pooled data in each partition. The details of the feature mapping technique in the VTL-BM framework is discussed in the next section.

### 3.4.2  Feature Mapping Using Multiple Background Models

As mentioned earlier in the previous section, feature mapping proposed in [13] compensates for channel mismatch between training and testing conditions. This section briefly outlines the feature mapping method. The entire training data of a root-UBM is partitioned into channel dependent datasets. Individual background models are derived by MAP adaptation of the root-UBM parameters using the channel dependent data. The mapping function is learned by examining the model parameters' shift and scale during MAP adaptation. All training and test data are transformed using the mapping function for an overall aggregation of multiple channel information into a single channel independent space. As discussed in Sect. 3.4.1, instead of channel dependent data, the VTL-wise partitioned data is used for MAP adaptation of a root-UBM [15]. The resultant VTL-BMs are independent of channel or gender information. The formal steps of the feature mapping procedure in the multiple background model framework are given as follows.

- Given an input feature vector X (training or testing), the most likely VTL-BM (say BM1) is detected based on its likelihood scores.
- The best Gaussian in BM1 is decoded for the vector X in the given utterance as given in Eq. (3.10)

$$i = \arg \max_{1 \leq j \leq M} \omega_j^{BM1} p_j^{BM1}(X) \qquad (3.10)$$

where $\omega_j^{BM1}$, $p_j^{BM1}$ are the weight and density of the $j$th Gaussian in BM1. M denotes the total number of Gaussian components in BM1.

- X is mapped to the VTL and channel independent vector Y by mean shift and variance scaling as given in Eq. (3.11).

$$Y = (X - \mu_i^{BM1}) \frac{\sigma_i^{CI}}{\sigma_i^{BM1}} + \mu_i^{CI} \qquad (3.11)$$

where CI represents the root-UBM. $\mu_i$ and $\sigma_i$ are the mean and diagonal covariance matrices for the $i$th Gaussian in both CI and BM1. The overall aim of the feature mapping technique is to transform $X \sim \mathcal{N}(\mu_i^{BM1}, \sigma_i^{BM1})$ to $Y \sim \mathcal{N}(\mu_i^{CI}, \sigma_i^{CI})$ where $\mathcal{N}(\mu_i, \sigma_i)$ denotes a Gaussian with mean $\mu_i$ and covariance $\sigma_i$.

**Fig. 3.4** Feature mapping in multiple background model framework

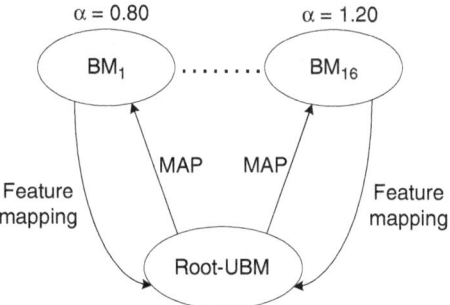

Figure 3.4 illustrates the feature mapping process. The feature-mapped training and test vectors are now used in the standard GMM-UBM framework. Target speaker GMMs are derived from the root-UBM using the mapped enrollment utterances. MAP adaptation of all the target speaker GMMs from a common root-UBM makes application of standard score normalization techniques convenient. Besides, a fast scoring based on the top N decoded Gaussians, can be applied during the evaluation stage of SV [3]. As mentioned in [13] feature mapping is applicable for multi-speaker speech recognition where each training utterance contains multiple types of speaker information. This is possible because the mapping function is constantly updated for each feature vector over a short-term frame according to Eqs. (3.10) and (3.11), respectively. The present work exploits this property to map non-overlapping segments of speech utterances, each of which is corrupted by a particular type of additive noise. Automatic updation of the mapping function over a short-term frame (20 ms) enables detection of noise changes occurring within the utterance. The resultant mapped features thus carry aggregate information from a noise-independent space.

## 3.5 Development of Speaker Verification Systems in Feature Mapping Framework

The experimental setup was identical to the one discussed in Sect. 3.3 earlier. All training and test utterances in each of the four artificially constructed non-homogenous noisy datasets (described in Sect. 3.3.1.2), were subjected to feature mapping using the 16 VTL-BMs and the root-UBM. Thus an additional four sets (one for each SNR) of feature mapped training and test utterances was obtained. Since the features were mapped frame-wise, it was assumed that no noise change occurred within a short-term frame of a noisy utterance. The UBM constructed earlier for the GMM-UBM framework (as discussed in Sect. 3.3.4) was used as the root-UBM. Four sets of target speaker GMMs were derived by MAP adaptation of the root-UBM using the training data in each of the four mapped datasets. Each set, consisting of 356 models (one for each enrolled speaker) was individually used

**Table 3.4** Summary of performance of the SV systems in Matched and Mismatched conditions at 0, 5, 7 and 10 dB SNRs

| SNR (dB) | Matched | | | | Mismatched | | | |
|---|---|---|---|---|---|---|---|---|
| | GMM-UBM | | Feature mapping | | GMM-UBM | | Feature mapping | |
| | EER (%) | MinDCF | EER (%) | MinDCF | EER (%) | MinDCF | EER (%) | MinDCF |
| 0 | 27.06 | 0.094 | 26.60 | 0.094 | 32.25 | 0.099 | 29.81 | 0.099 |
| 5 | 25.74 | 0.091 | 23.26 | 0.086 | 30.89 | 0.098 | 27.82 | 0.094 |
| 7 | 25.29 | 0.088 | 22.67 | 0.082 | 28.09 | 0.095 | 27.46 | 0.094 |
| 10 | 21.86 | 0.080 | 19.83 | 0.076 | 25.38 | 0.091 | 21.82 | 0.080 |

for evaluation. To demonstrate the overall utility of the proposed method towards environment adaptation, two sets of experiments were conducted in matched and mismatched conditions, respectively as discussed in Sect. 3.3.5. In the context of feature mapping, each of these conditions are described as

- Matched: Training and testing were both carried out in noisy environments. All training/enrollment and test utterances in each noisy dataset were subjected to feature mapping prior to model building and evaluation.
- Mismatched: The mismatch occurs due to clean training and noisy testing environments. The noisy test utterances in each of the four datasets were subjected to feature mapping prior to evaluation against clean speaker models developed using clean NIST training data.

The EER and MinDCF values were used as metrics for performance evaluation.

The performance of the SV systems in the GMM-UBM (baseline) and the feature mapping framework in matched and mismatched conditions have been summarized in Table 3.4.

As noticed from the tables, the performance of the feature mapping based SV systems is better than the baseline system in both the matched and mismatched conditions. The EER and MinDCF values are observed to decrease with the increase of SNR in each noisy dataset in all three conditions for both the SV systems. A consistent reduction of EER is noticed in case of the proposed systems across all SNRs. The performance improvement in terms of average EER reductions across all SNRs, is calculated. The proposed method shows a moderate improvement of 2.43 EER in mismatched condition. A more prominent improvement of 4.11 EER is noticed in case of matched conditions. The results indicate that the statistical similarities of noisy test utterances with noisy speaker models (Matched) are much improved when the all utterances are subjected to feature mapping. However, the mismatch due to differences in clean training and noisy test environments are not much improved despite feature mapping.

Figure 3.5 demonstrate the effect of feature mapping on the DET curves for the corresponding systems. Contrasting behavior of the broken lines (feature mapping) can be observed at each SNR. A distinct rotation can be observed in the black curves which represents performance under matched conditions. However the red curves, which represent performances in mismatched conditions in each plot are relatively more aligned at all operating points with the only exception in case of 10 dB SNR.

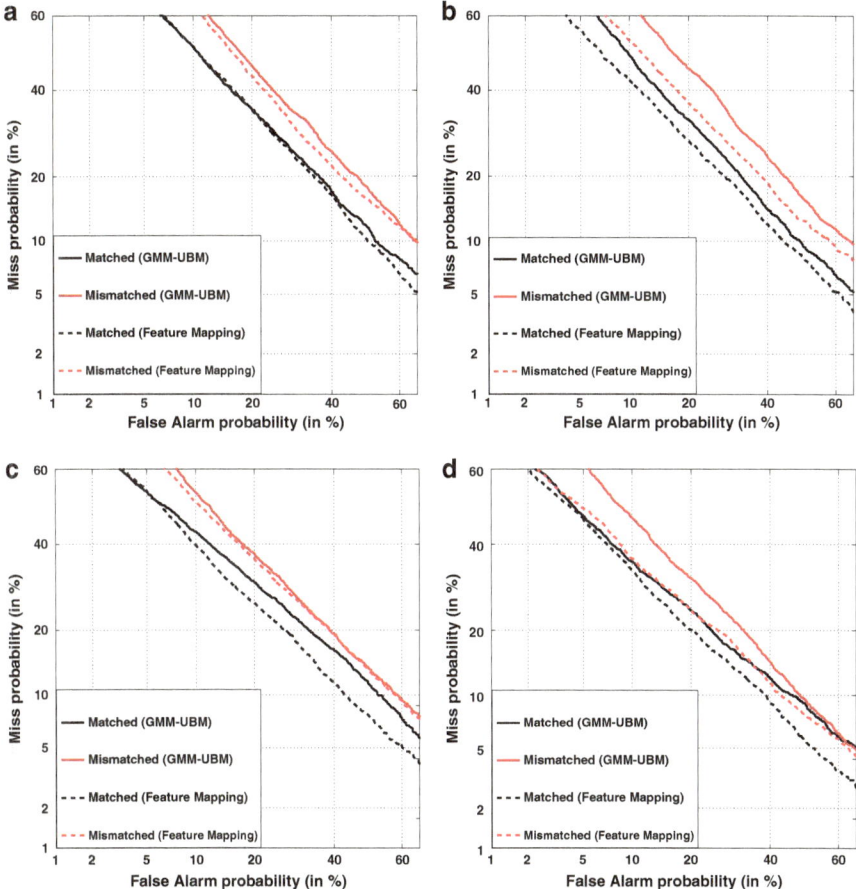

**Fig. 3.5** DET plots showing the effect of feature mapping in varying background environments at (**a**) 0 dB (**b**) 5 dB (**c**) 7 dB and (**d**) 10 dB SNR levels. The *broken black* and *red colored curves* in each plot indicate matched and mismatched conditions, respectively

The shift of the broken lines towards the origin also results in DCF reductions. The average MinDCF reductions across all SNRs are $3.75 \times 10^{-3}$ and $4 \times 10^{-3}$ for matched and mismatched conditions, respectively.

## 3.6  Summary

This chapter explored the GMM-UBM framework for speaker verification (SV) in various simulated noisy environments. Specifically, speech utterances corrupted in two types of background environments i.e., uniform and time-varying, were independently used for developing the SV systems. To understand the sensitivity of the GMM-UBM framework in practical situations, the SV systems were evaluated in clean, matched and two types of mismatched conditions. Results revealed severe performance degradation in matched and mismatched training-testing conditions as compared to the clean conditions. It was observed that performances in mismatched (type-II) condition were worse compared to the matched condition. Furthermore, results indicated overall inferior SV performance in the varying background environments in comparison to the uniform environments at equivalent SNRs.

To compensate the effect of varying background noise, a feature mapping technique was used for frame-wise transformation of noisy utterances. Results indicated moderate performance improvements in case of matched and mismatched conditions. The feature mapping effects were more prominent in case of matched conditions than in mismatched condition. However, since the goal of feature transformation is to address the issue in mismatched conditions, the tradeoff between the performance improvements and computational cost for developing the feature mapping framework is considered to be low. In the next chapter, advanced feature transformation techniques are explored. In the remaining chapters of the book, the GMM-UBM based SV systems shall be considered as baselines for performance comparison.

# References

1. T. Kinnunen, H. Li, An overview of text-independent speaker recognition: from features to supervectors. Speech Commun. **52**, 12–40 (2010)
2. D.A. Reynolds, R.C. Rose, Robust text-independent speaker identification using Gaussian mixture speaker models. IEEE Trans. Acoust. Speech Signal Process. **3**(1), 72–83 (1995)
3. D. Reynolds, T. Quatieri, R. Dunn, Speaker verification using adapted Gaussian mixture models. Digit. Signal Process. **10**(1), 19–41 (2000)
4. C.M. Bishop, *Pattern Recognition and Machine Learning* (Springer, New York, 2006)
5. J. Gauvain, C. Lee, Maximum a posteriori estimation for multivariate Gaussian mixture observations of Markov chains. IEEE Trans. Speech Audio Process. **2**(2), 291–298 (1994)
6. NIST-speaker recognition evaluations (1995), http://www.itl.nist.gov/iad/mig/tests/spk/
7. A. Varga, H.J. Steeneken, Assessment for automatic speech recognition: II. NOISEX-92: a database and an experiment to study the effect of additive noise on speech recognition systems. Speech Commun. **12**, 247–251 (1993)
8. S.R.M. Prasanna, G. Pradhan, Significance of vowel-like regions for speaker verification under degraded conditions. IEEE Trans. Audio Speech Lang. Process. **19**(8), 2552–2565 (2011)
9. A. Martin, G. Doddington, T. Kamm, M. Ordowski, M. Przybocki, The DET curve in assessment of detection task performance, in *Proceedings of the European Conference of Speech Communication Technology (EUROSPEECH '97)*, Rhodes, 1997, pp. 1895–1898
10. M.J.F. Gales, S.J. Young, Robust speech recognition in additive and convolutional noise using parallel model combination. Comput. Speech Lang. **9**, 289–307 (1995)
11. P.J. Moreno, B. Raj, R.M. Stern, A vector Taylor series approach for environment-independent speech recognition, in *Proceedings of IEEE International Conference on Acoustics, Speech and Signal Processing*, Atlanta, 1996, pp. 733–736
12. R. Teunen, B. Shahshahani, L. Heck, A model-based transformational approach to robust speaker recognition, in *Proceeding of the Annual Conference of the International Speech Communication Association (INTERSPEECH '00)*, Beijing, 2000, vol. 2, pp. 495–498

13. D.A. Reynolds, Channel robust speaker recognition via feature mapping, in *Proceedings of IEEE International Conference on Acoustics, Speech and Signal Processing (ICASSP '03)*, Hong Kong, 2003, pp. 53–56

14. W.Q. Zhang, Y. Shan, J. Liu, Multiple background models for speaker verification, in *Workshop on Speaker and Language Recognition (Odyssey)*, Brno, 2010

15. A. Sarkar, S. Umesh, Use of VTL-wise models in feature-mapping framework to achieve performance of multiple-background models in speaker verification, in *Proceedings of IEEE International Conference on Acoustics, Speech and Signal Processing (ICASSP '11)*, Prague, 2011

16. L. Lee, R. Rose, A frequency warping approach to speaker normalization. IEEE Trans. Speech Audio Process. **6**(1), 49–60 (1998)

17. S. Grashey, C. Geissler, Using a vocal tract length related parameter for speaker recognition, in *The Speaker and Language Recognition Workshop IEEE Odyssey*, San Juan, PR, USA, 2006

# Chapter 4
# Stochastic Feature Compensation for Robust Speaker Verification

**Abstract** This chapter explores the impact of standard stereo-based stochastic feature compensation (SFC) methods for robust speaker verification in uniform noisy environments. In this work, SFC using independent as well as joint probability models are explored for compensating the effect of noise. Integration of a SFC stage in the GMM-UBM framework is proposed for speaker verification evaluation under mismatched conditions.

The choice of features used for speaker recognition (SR) tasks is usually a tradeoff between accuracy, implementation costs and robustness. Short-term spectral or vocal tract features (e.g., MFCC) are the most extensively used for SR tasks due to their high speaker discriminative properties [1]. However, they are highly suscep-tible to noise-degradation and are therefore aided by compensation procedures in most SR applications [2,3]. The role of feature compensation was briefly introduced in Chap. 1. Despite the existence of inherent robust features, SR applications often prefer simple spectral features due to their ease of extraction. Such applications essentially require feature compensation methods for noise-robustness.

The discussion about the filtering-based feature compensation methods (e.g., CMS [4], RASTA [5]) in Chap. 2 revealed that they are specifically designed for cep-stral features and are commonly applied for suppressing channel effects. However, filtering is often inadequate for additive background environments where the log-spectral effect is ineffective. The application of model-based compensation schemes (e.g., SS [6], CDCN [7]) are likewise compromised due to the unavailability of a noise-model and high amount of training data.

The data-driven feature compensation methods offer a number of significant advantages compared to the other two categories. Firstly, they are independent of any analytical representation about the nature of the noise-corruption process. Secondly, they can better model the noise-effects due to their stochastic nature. Lastly, their performance is consistent across different environments. The only apparent drawback of applying these methods is the requirement of stereo data

K.S. Rao and S. Sarkar, *Robust Speaker Recognition in Noisy Environments*,
SpringerBriefs in Electrical and Computer Engineering,
DOI 10.1007/978-3-319-07130-5_4, © The Author(s) 2014

which can be interpreted as having a priori knowledge about the test environment. Despite such drawbacks, these techniques have been successfully used for far-field speech recognition tasks. To the best of the author's knowledge, the effect of these feature compensation methods have not been studied for robust speaker verification (SV) tasks. The application of standard stochastic feature compensation methods in a SV framework is proposed in this chapter. The significance of the proposed approach is demonstrated through a set of conducted experiments in simulated noisy environment.

The rest of the chapter is organized as follows. Section 4.1 gives a brief introduction to stochastic feature compensation, Sects. 4.2.1–4.3.2 provide detailed description of the feature compensation methods considered in the work [8], the proposed SV framework is discussed in Sect. 4.4 followed by a brief summary of the present work in Sect. 4.5.

## 4.1  Stochastic Feature Compensation (SFC)

Since accurate enumeration of the environmental effects on speech is a non-trivial task, a simplified form of speech signal degradation based on additive and convolutional channel noise is used in practice. Due to the random nature of noise, a given clean feature vector can generate different noisy feature vectors, and vice-versa, which causes an uncertainty. Conventionally, Gaussian Mixture Models (GMMs) are used to represent the cepstral distribution. The additive noise in general alters the distribution of mel frequency cepstral coefficients (MFCCs) by reducing the variance of each Gaussian component while the convolutional noise shifts the mean vectors.

Stochastic feature compensation (SFC) methods are independent of any mathematical structure of noise degradation. They model stereo training data using GMMs. Given a noisy test feature vector $y_t$, a minimum mean squared error (MMSE) criterion is used to estimate a clean vector $\hat{x}_t$ as follows

$$\hat{x}_t = E[x|y_t] = \int_X xp(x|y_t)\mathrm{d}x \tag{4.1}$$

where $x$ is a random variable representing clean feature vectors and $p(x|y_t)$ is the conditional probability distribution function (pdf) of $x$ given $y_t$. Depending on the nature of the feature compensation algorithm, the two broad approaches of deriving $p(x|y_t)$ can be categorized as (i) Independent probability modeling and (ii) Joint probability modeling. The independent probability modeling methods construct individual GMMs for clean and noisy data. The effect of noise is represented as additive terms to the mean vectors and covariance matrices of the GMMs. The conditional pdf is derived based on numerical approximations using the additive terms. Alternatively, joint probability models construct a single GMM using stacked noisy and clean feature vectors of the stereo data. This is followed by deriving an exact conditional pdf and estimation of clean speech vectors. Each of these methods are discussed in details in the following two sections

## 4.2  SFC Using Independent Probability Models

Figure 4.1 illustrates the independent probability model based SFC process. The main steps of the process can be outlined as follows

1. Firstly individual GMMs are built for clean vectors $X_t$ and noisy vectors $Y_t$ as follows

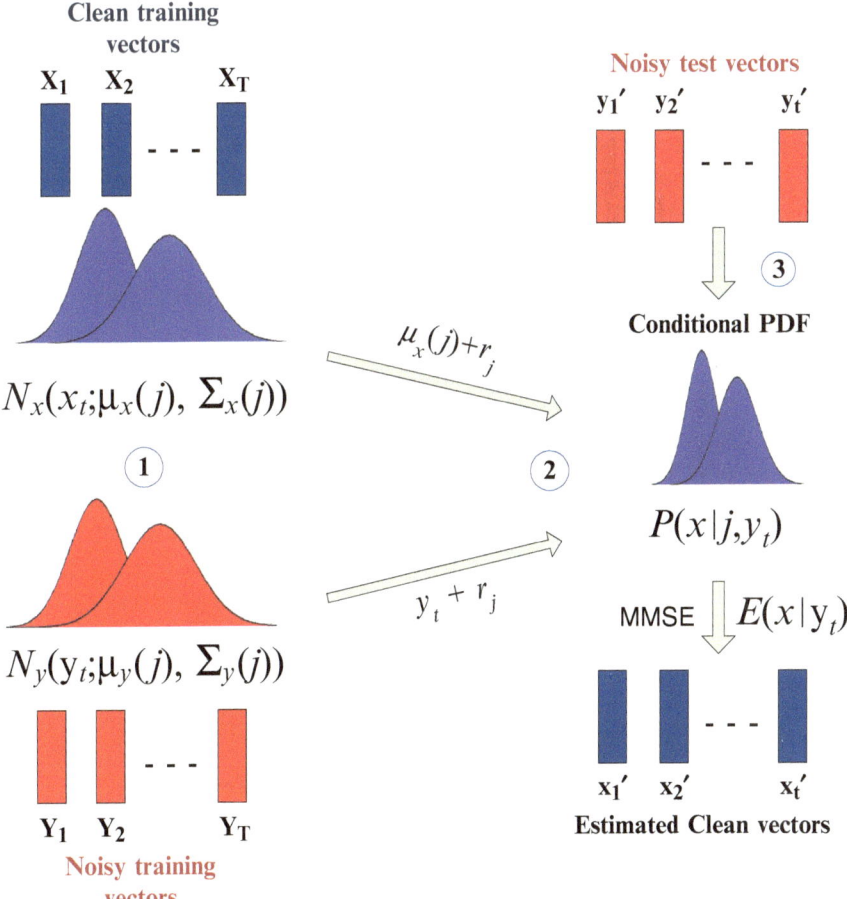

**Fig. 4.1** Stochastic feature compensation using independent probability models

$$p(x_t) = \sum_{j=1}^{M} w_x(j) \mathcal{N}_x(x_t; \mu_x(j), \Sigma_x(j)) \qquad (4.2)$$

$$p(y_t) = \sum_{j=1}^{M} w_y(j) \mathcal{N}_y(y_t; \mu_y(j), \Sigma_y(j)) \qquad (4.3)$$

where $w(j)$, $\mu(j)$ and $\Sigma(j)$ denotes the weight, mean vector and covariance matrix of the $j$th Gaussian component and $M$ is the total number of components.

2. The conditional pdf $p(x|j, y_t)$ is then approximated by means of additive factors $r_j$ to the clean or noisy training vectors. The values of the additive terms are determined my maximizing the likelihood of the training data.

3. Given a set of noisy test vectors, the equivalent set of clean vectors are estimated by MMSE

The GMM representations given by Eqs. (4.2) and (4.3), are used in the remaining chapter. In the following sections three standard independent probability model based SFC techniques used for robust speech recognition tasks are discussed briefly. Each of the methods differ in the way by which they derive $p(x|y_t)$ and thereby estimate $\hat{x}_t$. Detailed derivations of the additive terms and the MMSE estimator for each of these algorithms can be found in Appendix A.

### 4.2.1 Multivariate Gaussian-Based Cepstral Normalization (RATZ)

The RATZ algorithm [9], derives the required MMSE clean feature estimate in three stages. In the first stage, the clean feature vectors are used to train a GMM as in Eq. (4.2) using the standard Expectation Maximization (EM) algorithm. The second stage consists of estimating the statistics of the noise-degraded speech by applying appropriate correction vectors to the mean and covariance matrices of the clean speech pdf. The additive correction vectors, which model environmental effect are in turn estimated by maximizing the likelihood of the noisy feature vectors. Finally, given a noisy test feature vector, a MMSE estimate of clean speech is made using the correction vectors learned during the training phase. Given a sequence of $\mathbf{T}$ noisy MFCC vectors $Y = [y_1, y_2, \ldots y_T]$, the log-likelihood is given by

$$L(Y) = \log \prod_{t=1}^{T} p(y_t) = \sum_{t=1}^{T} \log \sum_{j=1}^{M} w_y(j) \mathcal{N}_y(y_t; \mu_y(j), \Sigma_y(j))$$

$$= \sum_{t=1}^{T} \log \sum_{j=1}^{M} w_y(j) \mathcal{N}_y(y_t; \mu_x(j) + r_j, \Sigma_x(j) + R_j)$$

$$(4.4)$$

where $r_j$ and $R_j$ are the correction vectors for the $j$th Gaussian component of the clean speech pdf. The complete set of unknown bias vectors is iteratively estimated by maximizing $L$ using an EM algorithm. Details of the EM algorithm have been outlined in Appendix A. The solutions obtained are given by the following equations

$$\hat{r}_j = \frac{\displaystyle\sum_{t=1}^{T} p(s_y(j)|y_t, \phi)(y_t - \mu_x(j))}{\displaystyle\sum_{t=1}^{T} p(s_y(j)|y_t, \phi)} \tag{4.5}$$

$$\hat{R}_j = \frac{\displaystyle\sum_{t=1}^{T} p(s_y(j)|y_t, \phi)\{(y_t - \mu_x(j) - \hat{r}_j)(y_t - \mu_x(j) - \hat{r}_j)^T - \Sigma_x(j)\}}{\displaystyle\sum_{t=1}^{T} p(s_y(j)|y_t, \phi)} \tag{4.6}$$

where $p(s_y(j)|y_t, \phi)$ is the posterior probability of the latent noisy GMM component $s_y(j)$ given $y_t$, $\phi = \{r_j, R_j\}$ is the set of model parameters and $T$ denotes matrix transpose. It was studied by Moreno et al. [9] that in case of stereo recordings, a one-one correspondence of the each Gaussian component of the noisy speech GMM and clean speech GMM can be established. This is done by assuming *posterior invariance* which states that the posterior probabilities of each GMM component with respect to a clean vector and its noisy equivalent vector are equal. This assumption, although less reliable in low SNR conditions suggest that each Gaussian undergoes the same shift and negligible compression. It gives a convenient approximation of $p(s_y(j)|y_t, \phi)$ as follows

$$p(s_y(j)|y_t, \phi) = \frac{p(s_y(j))p(y_t|s_y(j), \phi)}{\displaystyle\sum_{k=1}^{M} p(s_y(k))p(y_t|s_y(k), \phi)}$$

$$= \frac{p(s_x(j))p(x_t|s_x(j))}{\displaystyle\sum_{k=1}^{M} p(s_x(k))p(x_t|s_x(k))}$$

$$= \frac{w_x(j)\mathcal{N}_x(x_t; \mu_x(j), \Sigma_x(j))}{\displaystyle\sum_{j=1}^{M} w_x(j)\mathcal{N}_x(x_t; \mu_x(j), \Sigma_x(j))} \tag{4.7}$$

Given the above relation, Eqs. (4.5) and (4.6) can now be approximated as

$$\hat{r}_j = \frac{\displaystyle\sum_{t=1}^{T} p(s_x(j)|x_t)(y_t - x_t)}{\displaystyle\sum_{t=1}^{T} p(s_x(j)|x_t)} \tag{4.8}$$

$$\hat{R}_j = \frac{\sum\limits_{t=1}^{T} p(s_x(j)|x_t)\{(y_t - x_t - \hat{r}_j)(y_t - x_t - \hat{r}_j)^T - \Sigma_x(j)\}}{\sum\limits_{t=1}^{T} p(s_x(j)|x_t)} \qquad (4.9)$$

Since the above equations do not have $\phi$ in the right hand side, the solutions are non-iterative. The environmental effects on clean speech $x$ in MFCC domain are modeled as additive linear correction vectors $r(x)$. The MMSE estimate for clean speech $\hat{x}_t$ given a noisy test vector $y_t$ is calculated by Eq. (4.1). The conditional mean is solved using a numerical approximation as follows

$$\hat{x}_t = E[x|y_t] = y_t - \sum_{j=1}^{M} p(j|y_t)r_j \qquad (4.10)$$

## 4.2.2  Stereo Piece-Wise Linear Compensation for Environment (SPLICE)

The effectiveness of the RATZ algorithm depends on the posterior invariance assumption made in Eq. (4.7). However in low SNR conditions this assumption becomes unrealistic since the Gaussian pdfs of noisy speech are compressed in different amounts due to changes in its variance. As an alternative, the SPLICE algorithm proposed in [10] models the noisy feature space as given by the following equation

$$p(y_t) = \sum_{j=1}^{M} p(j)p(y_t|j) \qquad (4.11)$$

where $p(j)$ is the prior probability of the Gaussian component $j$ mathematically equivalent to the component weight $w_y(j)$ and $p(y|j)$ is the multivariate Gaussian $\mathcal{N}_y(y_t; \mu_y(j), \Sigma_y(j))$ as given in Eq. (4.3). A distinct advantage of SPLICE compared to other model-based feature enhancement techniques like Spectral Subtraction, is its consistent performance in non-stationary environments. Feature compensation using SPLICE is based on a two simple assumptions. Firstly, a clean MFCC vector $x_t$ generated by each discrete Gaussian component $j$ can be approximated in terms of its noisy counterpart $y_t$. This is often termed as piece-wise linear approximation. Secondly the conditional pdf of clean speech vectors given the noisy speech vectors and Gaussian component $j$ is also a multivariate Gaussian distribution. The mean of the resultant distribution is assumed to shifted by the corrective vector $r_j$ as follows

$$p(x|j, y_t) = \mathcal{N}_y(x; y_t + r_j, \Gamma_j) \qquad (4.12)$$

Estimation of the parameters $r_j$ and $\Gamma_j$ are based on maximum likelihood training similar to that of RATZ (Eq. 4.4) using an EM algorithm (outlined in Appendix A). The solutions are given by

$$\hat{r}_j = \frac{\sum_{t=1}^{T} p(j|y_t)(x_t - y_t)}{\sum_{t=1}^{T} p(j|y_t)} \tag{4.13}$$

$$\Gamma_j = \frac{\sum_{t=1}^{T} p(j|y_t)\{(x_t - y_t)(x_t - y_t)^T - \hat{r}_j\hat{r}_j^T\}}{\sum_{t=1}^{T} p(j|y_t)} \tag{4.14}$$

where $p(j|y_t)$ is the posterior probability of component $j$ given $y_t$

$$p(j|y_t) = \frac{p(j)p(y_t|j)}{\sum_{j=1}^{M} p(j)p(y_t|j)} \tag{4.15}$$

For stereo training data, the solution of Eqs. (4.13) and (4.14) are non-iterative. The MMSE estimate for clean speech from the noisy speech pdf is then given by

$$\hat{x}_t = E[x|y_t] = y_t + \sum_{j=1}^{M} p(j|y_t)r_j \tag{4.16}$$

The approximation of the mean of the conditional pdf in Eq. (4.12) using additive terms $r_j$ is often considered to be a limitation of the SPLICE framework. An accurate estimation of the conditional mean would require joint probability modeling of the clean and noisy vectors followed by estimating MLLR-type transforms [11]. Despite these drawbacks, SPLICE is commonly applied for pre-processing feature vectors in robust speech recognition tasks.

### 4.2.3 Multivariate Model Based Cepstral Normalization (MMCN)

The previous techniques discussed so far either models the clean feature space (e.g., RATZ) or the noisy feature space (e.g., SPLICE) using GMMs. A corrective bias vector for each GMM component is trained by weighing the difference between clean and noisy feature vector pairs with normalized posterior probabilities. However, in realistic situations when there are multiple types of environment in the noisy space, estimates based on single GMM posteriors might be erroneous. The Multi-Environment Model based LInear Normalization (MEMLIN) algorithm [12] aims to enhance performance accuracy by modeling both noisy and clean spaces in parallel. The noisy feature space is divided into several basic environments and modeled with individual GMM.

$$p_e(y_t) = \sum_{s_y^e=1}^{M} p(y_t|s_y^e)p(s_y^e) \tag{4.17}$$

where $s_y^e$ denotes the latent Gaussian component for the noisy GMM trained in environment indexed by $e$, $p_e(y_t|s_y^e)$ and $p_e(s_y^e)$ denote the Gaussian pdf for the $s_y^e$th component and its prior probability, respectively as shown below

$$p(y_t|s_y) = \mathcal{N}(y_t; \mu(s_y^e), \Sigma(s_y^e)) \tag{4.18}$$

$$p(s_y^e) = w_y^e \tag{4.19}$$

The clean feature space is modeled by a single GMM and has a similar structure as that of Eq. (4.2).

$$p(x_t) = \sum_{s_x=1}^{M} p(x_t|s_x)p(s_x) \tag{4.20}$$

The objective is to learn the difference between clean and noisy feature vectors associated with a pair of Gaussians (one for a clean model, and the other one for a noisy model), for each basic environment. The bias vector transformations are computed independently for each basic environment. Alike SPLICE, MEMLIN assumes that each clean feature vector $x_t$ is approximated by a linear function of the noisy feature vector $y_t$ and an additive bias vector $r(s_x, s_y^e)$. However unlike SPLICE, the additive vectors are now a function of both clean and noisy GMM components for a particular environment. The second assumption approximates the conditional pdf of $x$ given $y_t$ as a multivariate Gaussian with covariance matrix $\Sigma(s_x, s_y^e)$ and mean given by a linear transformation of the environment-dependent noisy vector, as follows

$$p(x|y_t, s_y^e, s_x) = \mathcal{N}(x|y_t - \sum_e p(e|y_t)r(s_x, s_y^e), \Sigma(s_x, s_y^e)) \tag{4.21}$$

where $p(e|y_t)$ and $r(s_x, s_y^e)$ are the posterior probability of environment $e$ given $y_t$ and the additive bias vector, respectively. The estimation of these factors are discussed briefly. The factor $p(e|y_t)$ is trained recursively as follows

$$p(e|y_t) = \beta p(e|y_{t-1}) + (1-\beta)\frac{p_e(y_{t-1})}{\sum_e p_e(y_{t-1})} \tag{4.22}$$

where $(0 \leq \beta \leq 1)$ is a constant and $p(e|y_0)$ is uniform across all environments. The $r(s_x, s_y^e)$ factor is obtained by maximizing the likelihood of noisy feature vector with respect to $r(s_x, s_y^e)$, using the standard EM algorithm. Given the stereo training data for environment $e$ which comprises the noisy vectors $Y_e = \{y_{t_e}\}_{t_e=1}^{T_e}$ and clean vectors $X_e = \{x_{t_e}\}_{t_e=1}^{T_e}$, the complete data log-likelihood of $Y_e$ is given by the following equation

$$L(Y_e) = \sum_{t_e=1}^{T_e} \log \sum_{s_y^e=1}^{M} p(s_y^e) \mathcal{N}_y(y_{t_e}; \mu(s_y^e) + r(s_x, s_y^e), \Sigma(s_x, s_y^e)) \qquad (4.23)$$

Maximizing the above equation with respect to $r(s_x, s_y^e)$ gives

$$r(s_x, s_y^e) = \frac{\sum\limits_{t_e=1}^{T_e} p(s_x|x_{t_e}) p(s_y^e|y_{t_e})(y_{t_e} - x_{t_e})}{\sum\limits_{t_e=1}^{T_e} p(s_x|x_{t_e}) p(s_y^e|y_{t_e})} \qquad (4.24)$$

where $p(s_x|x_{t_e})$ the posterior probability of Gaussian $s_x$ with respect to clean vector $x_t$. Similarly $p(s_y^e|y_{t_e})$ is the posterior probability of Gaussian $s_y^e$ with respect to noisy vector $y_t$. These can be easily calculated using Eqs. (4.20) and (4.17), respectively as follows

$$p(s_x|x_{t_e}) = \frac{p(x_{t_e}|s_x) p(s_x)}{\sum\limits_{s_x=1}^{M} p(x_{t_e}|s_x) p(s_x)} \qquad (4.25)$$

$$p(s_y^e|y_{t_e}) = \frac{p(y_t|s_y^e) p(s_y^e)}{\sum\limits_{s_y^e=1}^{M} p(y_t|s_y^e) p(s_y^e)} \qquad (4.26)$$

The resultant MMSE estimate $\hat{x}_t$ is computed as a weighted sum of all of the basic environment bias vector transformations.

$$\hat{x}_t = E[x|y_t] = y_t - \sum_{e} \sum_{s_y^e=1}^{M} \sum_{s_x=1}^{M} r(s_x, s_y^e) p(e|y_t) p(s_y^e|y_t) p(s_x|s_y^e, y_t, e)$$

$$(4.27)$$

The above equation introduces a new factor $p(s_x|s_y^e, y_t, e)$ known as cross probability model. It compensates for the mismatch that occurs when the Gaussian component $s_x$ associated with clean vector $x_t$ is different from the Gaussian component $s_y^e$ associated with corresponding noisy vector $y_t$. For simplicity the time dependency with $y_t$ is omitted, and the resultant factor $p(s_x|s_y^e, e)$ is estimated using relative frequency of occurrence. It is calculated as the ratio of the number of times the most probable pair of decoded Gaussians are $\{s_x, s_y^e\}$ and the number of times $s_y^e$ is decoded singly. The resultant form is as follows

$$p(s_x|s_y^e, e) = \frac{\sum\limits_{t_e=1}^{T_e} p(s_x|x_{t_e})p(s_y^e|y_{t_e})p(s_x)p(s_y^e)}{\sum\limits_{t_e=1}^{T_e}\sum\limits_{s_x=1}^{M} p(s_x|x_{t_e})p(s_y^e|y_{t_e})p(s_x)p(s_y^e)} \tag{4.28}$$

The single environment version of MEMLIN is often termed as Multivariate Model based Cepstral Normalization (MMCN). It can be easily deduced that in case of single environment, the variable $e$ can be omitted which simplifies most of the above equations. In such case the factor $p(e|y_t)$ can be entirely ignored. The scope of the present work is restricted to the single-environment version of MEMLIN i.e., MMCN.

## 4.3   SFC Using Joint Probability Models

The only apparent drawback of the independent probability model based SFC methods is the determination of the additive terms which may turn out be inaccurate in degraded environmental conditions. Alternatively, joint probability modeling can be used for feature compensation provided sufficient training data is available.

Figure 4.2 illustrates the independent probability model based SFC process. The main steps of the process can be outlined as follows

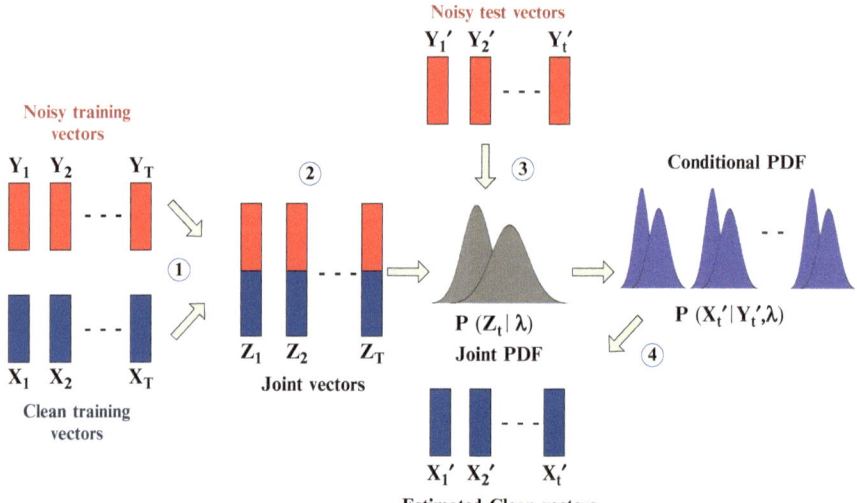

**Fig. 4.2** Stochastic feature compensation using joint probability models

1. The noisy and clean training vectors are concatenated to produce joint vectors (Z)
2. The joint vectors are modeled using a single GMM which represents the joint pdf
3. The conditional pdf is derived using parameters of the joint pdf.
4. Given a noisy test vector $Y_t'$, a clean vector $X_t'$ is obtained based on MMSE or maximum likelihood estimate (MLE).

Two standard joint probability model based SFC methods are discussed in the following sections

## 4.3.1  Stereo-Based Stochastic Mapping (SSM)

The main idea of the SSM algorithm [13] is to estimate the joint probability distribution of noisy and clean feature spaces instead of modeling them independently. This eliminates the need for training the hypothesized additive bias term '$r$' for each GMM component as employed by previous methods like SPLICE or MEMLIN. Unlike previous methods, concatenated pair of noisy and clean feature vectors are used as training data for GMM building. The desired transformation parameters are derived from the joint probability model (GMM) during the training phase. The improvement in performance accuracy is associated with a demand of larger amount of training data for estimating the model parameters in a higher dimensional space. The clean speech estimated during evaluation phase ($\hat{x}$), can be derived iteratively using MAP estimation or non-iteratively using the MMSE criterion. The scope of the present discussion is restricted to the MMSE version of the SSM for the ease of comparison with earlier methods. The details of the algorithm is described in the remaining part of this subsection.

As usual let's consider a pair of $d$ dimensional clean and noisy feature vector $x_t$ and $y_t$, respectively. A joint vector $z_t$ of dimension $2d$ is constructed as $z_t = [y_t^T, x_t^T]^T$. The joint vectors are modeled using a GMM $\lambda^{(z)}$ as follows

$$p(z_t) = \sum_{j=1}^{M} w_z(j) \mathcal{N}(z_t; \mu_z(j), \Sigma_z(j)) \tag{4.29}$$

where

$$\mu_z(j) = \begin{bmatrix} \mu_y(j) \\ \mu_x(j) \end{bmatrix}, \Sigma_z(j) = \begin{bmatrix} \Sigma_{yy}(j) & \Sigma_{yx}(j) \\ \Sigma_{xy}(j) & \Sigma_{xx}(j) \end{bmatrix} \tag{4.30}$$

This model is similar to those defined in Eqs. (4.2) and (4.3). The mean vector $\mu_z(j)$ for component $j$ is now a concatenation of individual mean vectors $\mu_y(j)$ and $\mu_x(j)$. The composition of the covariance matrix $\Sigma_z(j)$ can be similarly related. $\Sigma_{yy}(j)$ and $\Sigma_{xx}(j)$ are the covariance matrices for the $j$th component of the noisy and clean GMMs, respectively. Apart from these, $\Sigma_{yx}(j)$ and $\Sigma_{xy}(j)$ denote the cross-covariance matrices of $y$ and $x$ for the $j$th GMM component. The GMM is trained with the standard EM algorithm using the joint vectors $z$. The training stage essentially comprises deriving the model parameters by partitioning the matrices $\mu_z(j)$ and $\Sigma_z(j)$ as shown above. During the evaluation stage, the partitioned parameters are used to formulate the conditional pdf $p(x_t|y_t)$ required for the MMSE-based prediction of $\hat{x}_t$ as defined in Eq. (4.1). Unlike previous methods, mathematical derivations show that without any approximations the conditional pdf is another GMM where the mixture weights are posterior probabilities of each Gaussian component with respect to $y$ [14].

$$p(x_t|y_t, \lambda^{(z)}) = \sum_{j=1}^{M} p(j|y_t, \lambda^{(Z)}) p(x_t|y_t, j, \lambda^{(z)}) \tag{4.31}$$

where

$$p(j|y_t, \lambda^{(z)}) = \frac{w_y(j)\mathcal{N}(y_t; \mu_y(j), \Sigma_{yy}(j))}{\sum\limits_{j=1}^{M} w_y(j)\mathcal{N}(y_t; \mu_y(j), \Sigma_{yy}(j))} \tag{4.32}$$

$$p(x_t|y_t, j, \lambda^{(Z)}) = \mathcal{N}(x_t; E_x(j,t), D_x(j)) \tag{4.33}$$

The mean vector $E_x(j,t)$ and covariance matrix $D_x(j)$ of the $j$th Gaussian in the conditional pdf are defined as

$$E_x(j,t) = \mu_x(j) + \Sigma_{xy}(j)\Sigma_{yy}(j)^{-1}(y_t - \mu_y(j)) \tag{4.34}$$

$$D_x(j) = \Sigma_{xx}(j) - \Sigma_{xy}(j)\Sigma_{yy}(j)^{-1}\Sigma_{yx}(j) \tag{4.35}$$

Given a noisy test vector $y_t$, its equivalent clean estimate $\hat{x}_t$ can be then derived by the MMSE predictor as follows

$$\begin{aligned}
\hat{x}_t &= E[x_t|y_t] \\
&= \int_X x_t p(x_t|y_t, \lambda^{(z)})\mathrm{d}x_t \\
&= \int_X \sum_{j=1}^{M} x_t p(j|y_t, \lambda^{(Z)}) p(x_t|y_t, j, \lambda^{(z)})\mathrm{d}x_t \\
&= \sum_{j=1}^{M} p(j|y_t, \lambda^{(Z)}) E_x(j,t) \tag{4.36}
\end{aligned}$$

The principle of SSM is similar to SPLICE except for the joint probability distribution of noisy and clean feature spaces. In fact SPLICE with MMSE predictor reduces to its SSM counterpart if the cross-correlation of clean and noisy data is taken into account. SSM bears close resemblance to other model-based non-linear transformation methods like Constrained MLLR [15]. However the difference lies in the fact that the transformations in SSM are learned offline during the training phase while those in case of CMLLR, are done online during evaluation. A comparative study of SSM and other contemporary feature compensation methods can be found in [13].

### 4.3.2 Trajectory-Based Stochastic Mapping (TRAJMAP)

The MMSE estimator of SSM as discussed in Sect. 4.3.1 is a mixture of linear transforms weighted by the posterior probability of each GMM component. The parameters for the linear transform are derived from the joint distribution of both spaces. The approach is similar to any conventional GMM-based mapping techniques which has diverse applications [16]. However, a distinct drawback of such frame-wise mapping frameworks is that they fail to capture the correlation of features in the entire sequence. This results in inappropriate dynamic characteristics and an excessively smoothed spectra. The cepstral trajectory based GMM mapping (TRAJMAP) algorithm [17, 18] addressed this drawback by applying a Hidden Markov Model (HMM)-based parameter generation algorithm [19] with dynamic features, to the GMM-based mapping framework. Instead of individual frame-wise mapping, an entire sequence of frames (cepstral trajectory) is transformed in parallel. This approach had shown promising results for both noise-compensation [18] and voice conversion applications [17], in past. A fundamental assumption of the TRAJMAP algorithm is that despite noise corruption underlying spectral properties of a speaker remain preserved. The algorithm is used to learn a mapping function from a sequence of vectors in a speaker's noisy utterance to the corresponding sequence of clean vectors in the stereo training data. Mathematical details of the TRAJMAP transformation framework [17] is discussed in the remaining part of the section.

The cepstral vector trajectory is represented by a sequence of clean MFCC vectors $\mathbf{X}$ and noisy MFCC vectors $\mathbf{Y}$ where $\mathbf{X}$ and $\mathbf{Y}$ together constitute the stereo training data.

$$\mathbf{X} = [X_1^T, X_2^T, \ldots X_{\mathbf{T}}^T]^T \tag{4.37}$$

$$\mathbf{Y} = [Y_1^T, Y_2^T, \ldots Y_{\mathbf{T}}^T]^T \tag{4.38}$$

where $\mathbf{T}$ denotes the total number of vectors in the sequence. Individual vectors of each sequence are a concatenation of the static MFCC, its delta and acceleration coefficients. Each vector in the above sequence are $3d$ dimensional considering static MFCC vectors of $d$ dimension,

$$X_t = [x_t^T, \Delta x_t^T, \Delta^2 x_t^T]^T \tag{4.39}$$

$$Y_t = [y_1^T, \Delta y_t^T, \Delta^2 y_t^T]^T \tag{4.40}$$

The GMM $\lambda^{(Z)}$ of the joint pdf $p(Z_t|\lambda^{(Z)})$ is trained by a concatenated pair of clean and noisy vector $(Z_t)$ from the stereo training data where $Z_t = [Y_t^T, X_t^T]^T$. The aim to map the noisy MFCC trajectory $\mathbf{Y}$ to its clean counterpart $\mathbf{X}$. This is achieved by maximizing the following likelihood function

$$p(\mathbf{X}|\mathbf{Y}, \lambda^{(Z)}) = \sum_{\mathbf{j}} p(\mathbf{j}|\mathbf{Y}, \lambda^{(Z)}) p(\mathbf{X}|\mathbf{Y}, \mathbf{j}, \lambda^{(Z)})$$

$$= \prod_{t=1}^{T} \sum_{j=1}^{M} p(j|Y_t, \lambda^{(Z)}) p(X_t|Y_t, j, \lambda^{(Z)}) \qquad (4.41)$$

where $\mathbf{j} = \{j_1, j_2 \ldots j_T\}$ is a mixture component sequence. The conditional pdf at each frame is modeled as a GMM. At frame $t$, the $j$th mixture component weight $p(j|Y_t, \lambda^{(Z)})$ and the $j$th conditional probability distribution $p(X_t|Y_t, j, \lambda^{(Z)})$ are given by the following expressions

$$p(j|Y_t, \lambda^{(Z)}) = \frac{w_j^Y \mathcal{N}(Y_t; \mu_j^Y, \Sigma_j^{YY})}{\sum_{j=1}^{M} w_j^Y \mathcal{N}(Y_t; \mu_j^Y, \Sigma_j^{YY})} \qquad (4.42)$$

$$p(X_t|Y_t, j, \lambda^{(Z)}) = \mathcal{N}(X_t; E_{j,t}^X, D_j^X) \qquad (4.43)$$

where

$$E_{j,t}^X = \mu_j^X + \Sigma_j^{XY}(\Sigma_j^{YY})^{-1}(Y_t - \mu_j^Y) \qquad (4.44)$$

$$D_j^X = \Sigma_j^{XX} - \Sigma_j^{XY}(\Sigma_j^{YY})^{-1}\Sigma_j^{YX} \qquad (4.45)$$

The notations for conditional mean and conditional covariance used in Eqs. (4.44) and (4.45) are similar to the ones discussed earlier in Sect. 4.3.1.

The task is to estimate a sequence of clean vectors $\hat{X}$ from the entire sequence of noisy feature vectors $\mathbf{Y}$. This is achieved in two stages. In the first stage, a HMM-based parameter generation algorithm [19] is used to convert $\mathbf{Y}$ to the static MFCC parameters $\hat{x}$. In the next stage, the delta and acceleration coefficients are derived from each static MFCC vector of $\hat{x}$ and concatenated with itself to obtain the resultant sequence $\hat{X}$. In contrast to the MMSE-based methods, the derivation of $\hat{x}$ is based on a maximum likelihood estimate (MLE) as follows

$$\hat{x} = \arg\max_{x} p(\mathbf{X}|\mathbf{Y}, \lambda^{(Z)}) \qquad (4.46)$$

where $\hat{x} = [\hat{x}_1^T, \hat{x}_2^T, \ldots, \hat{x}_T^T]$ is the sequence of estimated static feature vectors. A matrix $\mathbf{W}$ of dimension $3d\mathbf{T} \times d\mathbf{T}$ is defined such that it converts the static sequence $\hat{x}$ to the expanded sequence $\hat{X}$ as follows

$$\hat{X} = \mathbf{W}\hat{x} \qquad (4.47)$$

where $\hat{X}$ is the sequence of denoised MFCC vectors with dynamic (delta and acceleration) co-efficients as defined in Eqs. (4.38) and (4.40). The composition of the matrix $\mathbf{W}$ is discussed as follows

$$\mathbf{W} = [W_1, W_2, \ldots W_t, \ldots W_\mathbf{T}]^T \otimes I_{DXD} \tag{4.48}$$

$$W_t = [\mathbf{w_t^{(0)}}, \mathbf{w_t^{(1)}}, \mathbf{w_t^{(2)}}] \quad t = 1, 2, \ldots \mathbf{T} \tag{4.49}$$

$$\mathbf{w_t^{(n)}} = [\overset{1st}{0}, \ldots, 0, \overset{(t-L_-^{(n)})th}{w^{(n)}(-L_-^{(n)})}, \ldots, \overset{(t+L_+^{(n)})th}{w^{(n)}(L_+^{(n)})}, \ldots, \overset{(t)th}{w^{(n)}(0)}, \ldots, \overset{\mathbf{T}\text{-th}}{0}]^T \quad n = 0, 1, 2 \tag{4.50}$$

In Eq. (4.48), each submatrix $W_t$ is of size $\mathbf{T} \times 3$ and '$\otimes$' denotes the Kronecker product. In Eq. (4.50), $w^{(n)}(\tau)$ denotes the weights required for calculating the $\Delta^n$ MFCC coefficient for the $(t + \tau)$th time frame. $\tau$ varies in a frame span of $[-L_-^{(n)}, L_+^{(n)}]$ as shown in the following equations ($L_+^{(0)} = L_-^{(0)} = 0$ and $w^{(0)}(0) = 1$)

$$\Delta x_t = \sum_{\tau=-L_-^{(1)}}^{L_+^{(1)}} w^{(1)}(\tau) x_{t+\tau} \tag{4.51}$$

$$\Delta^2 x_t = \sum_{\tau=-L_-^{(2)}}^{L_+^{(2)}} w^{(2)}(\tau) x_{t+\tau} \tag{4.52}$$

The maximum likelihood estimate in Eq. (4.46) is solved by an EM algorithm which iteratively maximizes an auxillary function with respect to $\hat{x}$ as follows

$$Q(\mathbf{X}, \hat{\mathbf{X}}) = \sum_j p(\mathbf{j}|\mathbf{Y}, \mathbf{X}, \lambda^{(Z)}) \log(p(\hat{\mathbf{X}}, \mathbf{j}|\mathbf{Y}, \lambda^{(Z)})) \tag{4.53}$$

The sequence of vector $\hat{x}$ obtained as a solution of Eq. (4.53) is given by

$$\hat{x} = (\mathbf{W}^T \overline{(\mathbf{D^X})^{-1}} \mathbf{W})^{-1} \mathbf{W}^T \overline{(\mathbf{D^X})^{-1} \mathbf{E^X}} \tag{4.54}$$

where

$$\overline{(\mathbf{D^X})^{-1}} = diag[\overline{(D_1^X)^{-1}}, \overline{(D_2^X)^{-1}}, \ldots, \overline{(D_t^X)^{-1}}, \ldots, \overline{(D_\mathbf{T}^X)^{-1}}] \tag{4.55}$$

$$\overline{(\mathbf{D^X})^{-1} \mathbf{E^X}} = [\overline{(D_1^X)^{-1} E_1^X}^T, \overline{(D_2^X)^{-1} E_2^X}^T, \ldots, \overline{(D_t^X)^{-1} E_t^X}^T, \ldots, \overline{(D_\mathbf{T}^X)^{-1} E_\mathbf{T}^X}^T]^T \tag{4.56}$$

$\overline{(\mathbf{D^X})^{-1}}$ in Eq. (4.55) is a block diagonal matrix of size $3d\,\mathbf{T} \times 3d\,\mathbf{T}$ while $\overline{(\mathbf{D^X})^{-1}\mathbf{E^X}}$ in Eq. (4.56) is a vector of size $3d\,\mathbf{T} \times 1$. The individual constituents of the matrices i.e., $\overline{(D_t^X)^{-1}}$ and $\overline{(D_t^X)^{-1}E_t^X}$ are given by

$$\overline{(D_t^X)^{-1}} = \sum_{j=1}^{M} \lambda_{j,t}(D_j^X)^{-1} \tag{4.57}$$

$$\overline{(D_t^X)^{-1}E_t^X} = \sum_{j=1}^{M} \lambda_{j,t}(D_j^X)^{-1}E_{j,t}^X \tag{4.58}$$

$$\lambda_{j,t} = p(j|Y_t, X_t, \lambda^{(Z)}) \tag{4.59}$$

Detailed derivation of Eq. (4.54) is provided in Appendix A. The solution $\hat{x}$ is only a sequence of static MFCC vectors i.e., a vector of size $d\,\mathbf{T} \times 1$. The full sequence with delta and acceleration coefficients appended with the resultant vector can be obtained by a simple linear operation $\mathbf{W}\hat{x}$.

## 4.4  Development of Proposed SV Systems

All experiments are carried out in the NIST-2003-SRE database [20] introduced in Chap. 3. The data consists of single training utterances of approximately 2 min length from each of 356 enrolled speakers and 3,500 test utterances (approximately 10–15 s each) for evaluation. The purpose of present work is to address the issue of speaker verification in mismatched condition where a speaker enrolls in a clean environment whereas during verification his/her speech is corrupted by background noise. However stereo-data based techniques as described in Sect. 4.1 require simultaneous recording of a speaker's training data over two channels i.e., one in clean condition and the other in a noisy environment. Due to unavailability of such data, the noisy utterances used in the present work were simulated by corrupting the clean speech utterances of the NIST-SRE-2003 by different types of additive noises. The approach is motivated by synthetic generation of stereo-data as described in [21]. The standard GMM-UBM framework was used for speaker verification [22]. Figure 4.3 shows the block diagram of the feature compensation process in a GMM-UBM based speaker verification system. The various stages of the SV system development are discussed in the following sections.

### 4.4.1  Simulation of Background Environment

Four additive noises (i.e., car, factory, pink and white) collected from the NOISEX-92 database [23] were used for representing unique background environments. The speech segment from each of the 356 enrolled speakers was degraded by adding a

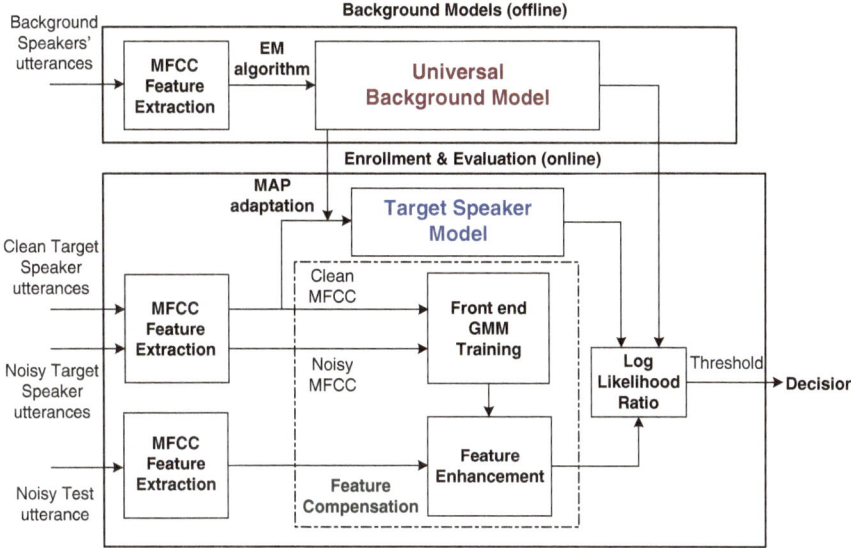

**Fig. 4.3** Block diagram of the feature compensation process in the speaker verification system

specific type of noise at 0 and 5 dB SNRs, respectively. The noise level was scaled to maintain the desired SNRs of the reconstructed speech segments. Eight different sets of noisy training utterances were obtained (one for each noise at a particular SNR). The default training set of the NIST-SRE-2003 was used as the clean recordings.

All test utterances were similarly reconstructed by noise addition at the two SNRs. Each set of noisy utterances were used for the following sets of experiments.

1. <u>Mismatched Condition</u>: The noisy test utterances were evaluated against speaker models built from clean training data.
2. <u>Matched Condition</u>: The noisy test utterances were evaluated against speaker models built from noisy training data.
3. <u>Feature compensated</u>: The noisy test utterances were subjected to feature enhancement prior to evaluation against clean speaker models. Each of the four feature compensation techniques discussed in Sect. 4.1 were compared with the above two conditions and the proposed method, on the basis of their performance.

The simulated stereo training data was used for front-end GMM training as discussed later in Sect. 4.4.3. For comparing relative improvements in performance accuracy produced by the various feature compensation schemes, the SV systems under mismatched conditions have been considered as a baseline.

## 4.4.2 Feature Extraction and Speaker Modeling

The feature extraction and speaker modeling process are identical to that used in the GMM-UBM framework described in Chap. 3. Standard MFCC coefficients were used as features. After pre-emphasis and an energy-based voiced activity detection, 39-dimensional feature vectors (consisting of 13 MFCCs $+ \Delta + \Delta\Delta$ excluding $C_0$) derived from a 26 channel mel-scaled filterbank, were extracted from speech frames of 20 ms with a frame-overlap of 10 ms. All feature vectors were subjected to cepstral mean subtraction followed by cepstral variance normalization. The resultant distribution was scaled to zero mean and unit variance. In the remaining part of the chapter, the MFCC feature vectors extracted from the noisy training data and its clean counterpart are referred as 'noisy vectors' and 'clean vectors', respectively.

Acoustic modeling using the standard GMM-UBM framework was performed in two stages i.e., construction of a Universal Background Model (UBM) and the target speaker models. Twenty hours of speech collected from the SwitchBoard II corpus was used to construct a 1,024-component GMM offline using 200 iterations of the EM algorithm. The target speaker models (GMMs) were derived by MAP adaptation of the UBM using each enrolled speaker's training data. The process was repeated twice, once each for the clean and noise-degraded speech of the stereo training data. The clean speaker models were used for evaluation in the mismatched condition as well as the feature compensated conditions.

## 4.4.3 Feature Compensation

The two basic stages of the feature compensation process are discussed below.

- **Front-end GMM Training**: The stereo training data corresponding to each speaker was used for building speaker-specific front-end GMMs prior to feature enhancement. For RATZ, SPLICE and MMCN, a pair of 8-component GMMs (clean and noisy) with diagonal covariance matrices were constructed for each speaker using the standard EM algorithm.

  For SSM and TRAJMAP, individual pairs of noisy and clean MFCC vectors in the aligned sequence were first concatenated to create a single sequence of 78-dimensional MFCC vectors. The joint vectors were used to build speaker specific 8-component GMMs with full covariance matrices. The number of components for the GMMs were empirically determined according to the available training data. However in practical applications without training data constraints, higher number of components can be explored. The conditional GMM parameters required for SSM and TRAJMAP were derived using Eqs. (4.34), (4.35) and (4.44), (4.45), respectively.

- **Feature Enhancement**: Each noisy test feature vector was transformed using the front-end GMM parameters of each of the 11 target speaker models specified for the evaluation phase of the NIST-2003 primary task. In contrast to the actual evaluation process, each of the 11 transformed vectors were scored against the corresponding speaker model and the UBM.

The corrective bias vectors of the mean and covariance terms for RATZ were estimated using Eqs. (4.8) and (4.9), non-iteratively. This was followed by the MMSE predicted value given by Eq. (4.10). Only the noisy front-end GMMs trained as in Eq. (4.3) were used to estimate the bias vectors for SPLICE as given by Eqs. (4.13) and (4.14). This was followed by the MMSE estimate given by Eq. (4.16). The simplified single environment version of MEMLIN i.e., MMCN was used for feature enhancement. The posterior probability factor for each environment given by Eq. (4.22) could thus be entirely omitted. The cross probability model (Eq. 4.28) and the MMSE predictor (Eq. 4.27) were likewise simplified. MMSE estimates for SSM and the MLE estimate for TRAJMAP were calculated using Eqs. (4.36) and (4.54), respectively. The static MFCCs obtained from TRAJMAP were concatenated with the delta and acceleration coefficients to yield the resultant 39-dimensional vector.

### 4.4.4  Effect of Feature Compensation in Cepstral Domain

Effectiveness of the stochastic feature compensation methods is demonstrated by a set of plots which highlight some characteristics of the transformed and distorted MFCC features. Since the lower order MFCC coefficients represent the broad spectral shape, the first MFCC coefficient has been considered without loss of generality for demonstrating the impact of feature compensation. Figure 4.4a shows the histogram of the first MFCC coefficients of an arbitrary test speech utterance from the NIST-2003-SRE and its equivalent noisy signal obtained by white noise addition at 0 dB SNR. Figure 4.4b–f shows the effect of enhancing the noisy utterance by applying various feature compensation algorithms. Since the feature vectors were mean and variance normalized, both the distributions are centered around zero. However the area under the overlapping region of the curves is a measure of accuracy in the conversion process. A fully overlapped curve would suggest the ideal situation of perfect conversion. The distortion caused by noise statistics can be observed in Fig. 4.4a in which the peak of noisy distribution is significantly skewed towards the left. The skewness shows a gradual reduction after the application of feature compensation algorithms. The shape of the transformed (compensated) distributions is similarly affected by noise addition. The simple noisy distribution shows arbitrary changes in the spectral shape as seen in several regions of the curve. The change in spectral shape is negligible in case of RATZ with minor differences at the peak region. The change in the noisy distribution shows more prominence in case of SPLICE and MMCN. A spectral smoothening effect can be observed at the peak regions for the SPLICE and MMCN-compensated distributions, respectively with slightly more overlap in case of the former. The SSM and TRAJMAP compensated

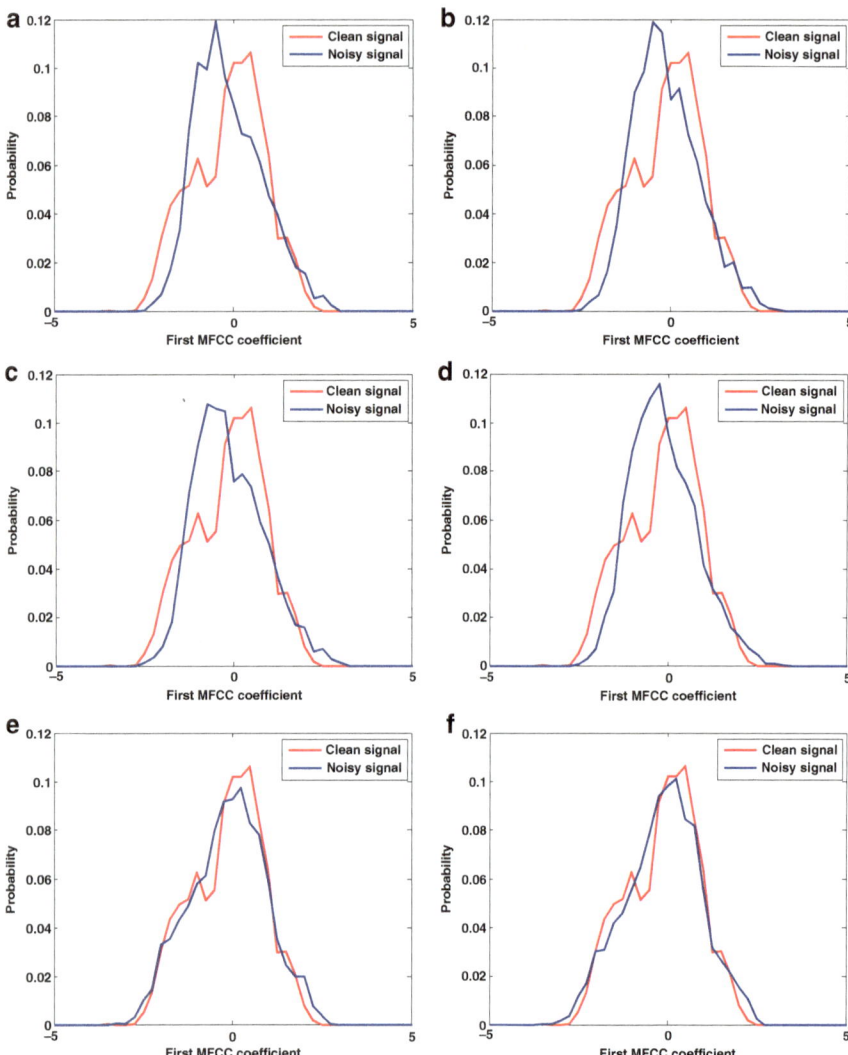

**Fig. 4.4** Histograms of the first MFCC coefficient of a clean test speech signal (*red*) and the same signal contaminated with white noise at 0 dB (*blue*) (**a**) without feature compensation and with feature compensation using (**b**) RATZ (**c**) SPLICE (**d**) MMCN (**e**) SSM and (**f**) TRAJMAP

distributions shows comparatively higher resemblances with the clean distribution. The significant increase in the overlapping area of the histograms is apparent. The changes are also reflected on the spectral shape which shows that the transformed distribution captures minute similarities at the peak region.

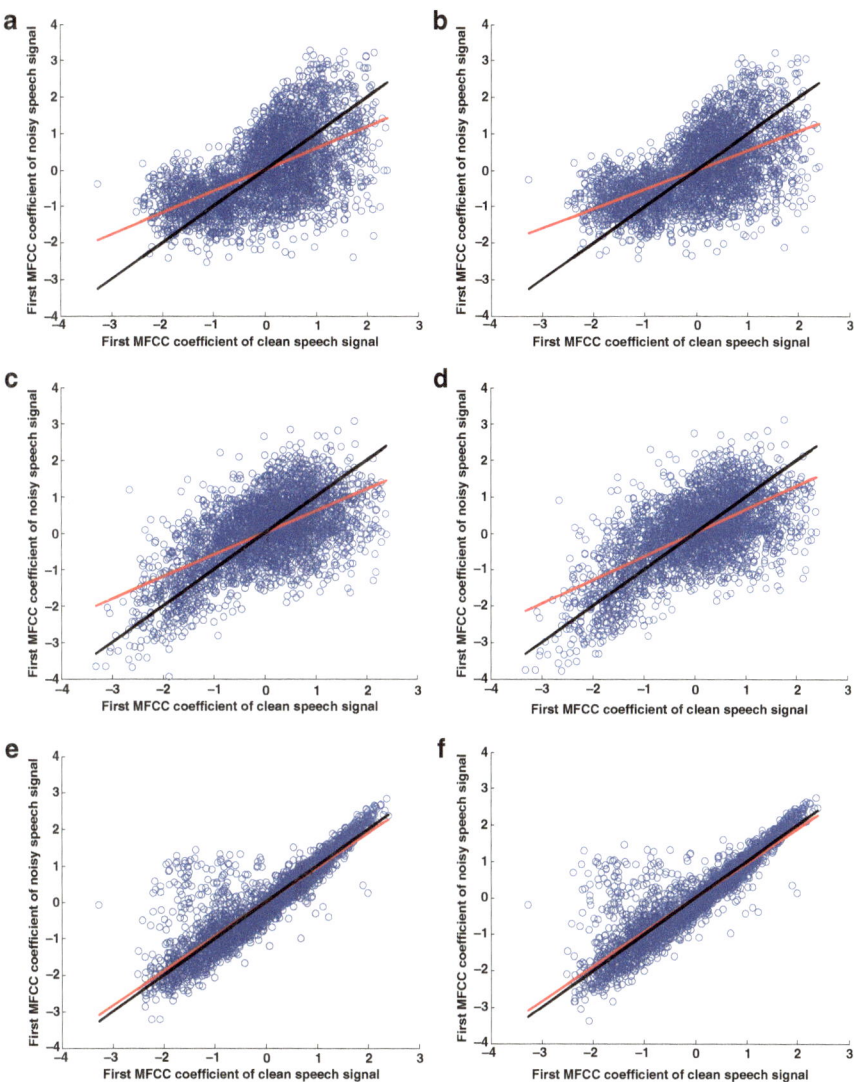

**Fig. 4.5** Scatterplots between the first MFCC coefficient of non-silence frames of a clean test speech signal (x-axis) and the same signal contaminated with white noise at 0 dB (y-axis) (**a**) without feature compensation and with feature compensation using (**b**) RATZ (**c**) SPLICE (**d**) MMCN (**e**) SSM and (**f**) TRAJMAP. The $(x = y)$ line is colored *black* while the line of best fit is colored *red*

Figure 4.5a–f shows the scatter plots between the first MFCC coefficients ($C_1$) of the given test utterance (x-axis) and its white noise corrupted equivalent (y-axis). The $C_1$s extracted from non-silence frames of the test utterance are represented by blue circles. The black line represents the ideal condition of perfect feature

transformation ($x = y$). The red line is a first order polynomial of the clean
feature vectors which best fits the noisy feature vectors in a least square sense.
The imperfections in the transformation process can be inferred from the deviation
between the two lines in a figure. The distortion of the cepstral distribution due
to the addition of white noise is apparent from Fig. 4.5a in which the two lines
are significantly deviated from each other. The spread of the data (blue dots)
across the black line is a measure of the covariance of the clean and noisy data.
Significant changes in the scatter plots can be observed after the application of the
feature compensation algorithms. The increased covariance of data is noticed in
case of SPLICE and MMCN where the deviation between the red and black lines is
relatively lower compared to RATZ. SSM and TRAJMAP shows the best fit in terms
of covariance of the given data with the latter showing marginal improvements over
the former. Despite outliers most of the data points are considerably aligned along
the line of best fit with very little noticeable deviation.

## 4.4.5   Performance Evaluation

All experiments were performed in mismatched, matched and compensated conditions each of which has been discussed in Sect. 4.4.1. The NIST-2003 primary task
was carried out in which each noisy test utterance was evaluated against 11 target
speaker models (GMMs). The equal error rate (EER) and minimum DCF (MinDCF)
values were used as metrics for performance evaluation.

Figures 4.6 and 4.7 show the DET curves of the SV systems in various conditions,
with background noise at 0 and 5 dB SNRs, respectively. The summary of the
performance of SV systems in different noisy background is shown in Table 4.1.
A quick observation reveals a general order of precedence of the SV performance
accuracy in terms of EER values i.e., mismatched, RATZ, matched, MMCN,
SPLICE, SSM and TRAJMAP. The pattern is also valid for the MinDCF values
except for the fact that they often do not show a monotonic decrement across the
various methods. The only exception to this order is seen in case of car noise at 0 and
5 dB SNRs. The mismatched condition expectedly shows the worst case scenario
in every noisy environments with an average EER of 29.93 % across all of them
for both SNRs. This is in conformity with the known fact that noise degradation
causes arbitrary changes in the clean feature distributions due to which noisy test
utterances yield poor scores during the pattern matching stage. The performance
of the RATZ compensation scheme shows minor improvement over the baseline
(mismatch) with an average decrement of 3.07 % EER. Interestingly, the matched
condition in most cases outperform RATZ. A possible explanation to this behavior
is the invalidity of the posterior invariance assumption in low SNR conditions as
discussed in Sect. 4.2.1. The effect of feature normalization using posterior probability of the noisy MFCC vectors with respect to clean Gaussian components has
other interesting implications. As discussed in Sect. 4.2.3, the MEMLIN algorithm
(multienvironment version of MMCN), uses both noisy and clean GMMs as inputs,

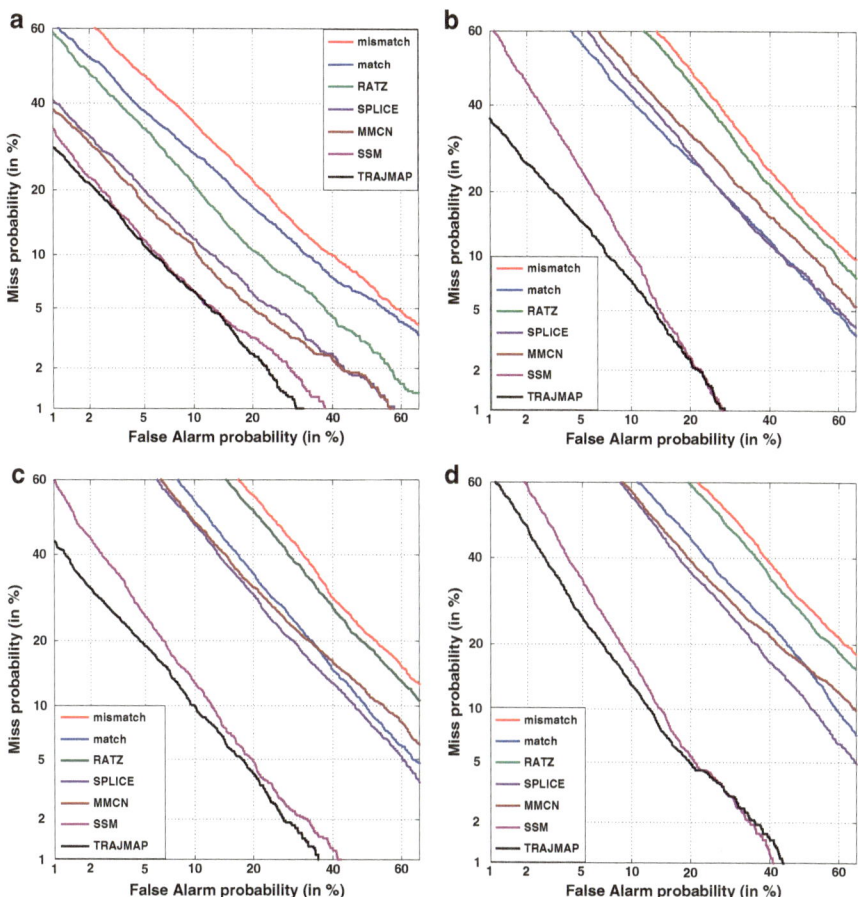

**Fig. 4.6** DET plots for the SV systems developed using the features derived from SFC methods in uniform background environment containing (**a**) car noise (**b**) factory noise (**c**) pink noise and (**d**) white noise at 0 dB SNR

thus incorporating both types of posterior probabilities in the final transformation. However, contrary to known facts, the SPLICE algorithm performs moderately better than the MMCN algorithm with an average improvement of 1.62 % EER for factory, pink and white background environments. The improvement is consistent in case of MinDCF values and more pronounced in case of factory, pink and white noises. There are two possible justifications to this phenomenon. Firstly, the inclusion of the inaccurate clean Gaussian posteriors in estimating the corrective vectors and secondly, an oversimplified cross probability model which excludes the environment factor '$e$' from the final transformations, as discussed in Sect. 4.2.3. It is interesting to note that this effect is in conformity with the anomalous behavior of the SV performances observed in case of car noise. Unlike the other background

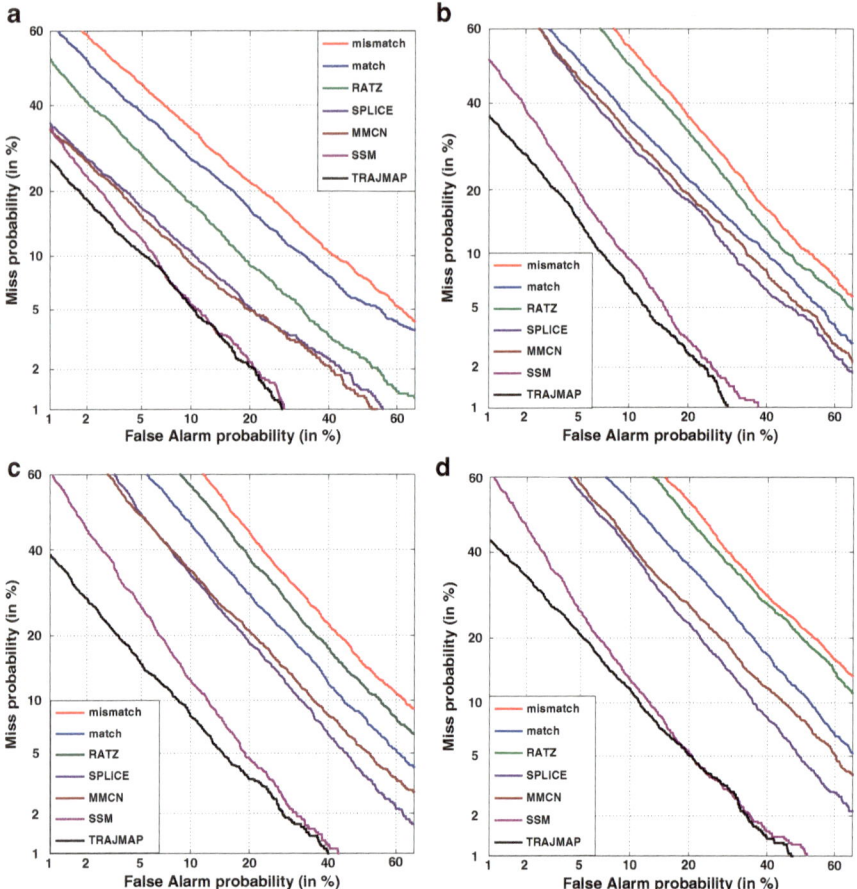

**Fig. 4.7** DET plots for the SV systems developed using the features derived from SFC methods in uniform background environment containing (**a**) car noise (**b**) factory noise (**c**) pink noise and (**d**) white noise at 5 dB SNR

noises, in case of car environment, it is observed that the performance in mismatched condition is only slightly worse than that of the matched one with an average drop of 2.67 % EER across both SNRs. In this case, the positive effect of clean Gaussian components in normalization, is also reflected by the considerable improved SV performances of RATZ and MMCN in comparison to the matched condition and SPLICE, respectively.

The SSM and TRAJMAP shows a significant improvement in performance compared to the other algorithms with a large margin of difference in terms of EER and MinDCF. In comparison to SPLICE, an average EER reduction as high as 9.37 and 10.32 % is obtained for SSM and TRAJMAP, respectively. The improvement is consistent even in the case of the anomalous car noise in which TRAJMAP is seen

**Table 4.1** Summary of performance of the SV systems developed using the features derived from SFC methods in uniform background environments

| SNR (dB) | Methods | Car | | Factory | | Pink | | White | |
|---|---|---|---|---|---|---|---|---|---|
| | | EER (%) | MinDCF | EER (%) | MinDCF | EER (%) | MinDCF | EER (%) | MinDCF |
| 0 | Mismatch | 20.55 | 0.079 | 32.16 | 0.099 | 35.05 | 0.099 | 39.02 | 0.099 |
| | Matched | 18.04 | 0.071 | 23.17 | 0.097 | 26.65 | 0.092 | 30.98 | 0.097 |
| | RATZ | 14.32 | 0.066 | 30.44 | 0.099 | 33.42 | 0.098 | 37.62 | 0.099 |
| | SPLICE | 10.75 | 0.049 | 23.26 | 0.085 | 23.87 | 0.096 | 27.32 | 0.098 |
| | MMCN | 9.94 | 0.048 | 26.02 | 0.089 | 25.47 | 0.095 | 28.91 | 0.096 |
| | SSM | 7.59 | 0.041 | 9.49 | 0.058 | 11.02 | 0.063 | 12.19 | 0.078 |
| | TRAJMAP | 7.45 | 0.039 | 8.45 | 0.045 | 9.71 | 0.051 | 11.07 | 0.066 |
| 5 | Mismatch | 20.95 | 0.077 | 27.15 | 0.098 | 30.39 | 0.097 | 34.16 | 0.099 |
| | Matched | 18.11 | 0.071 | 20.96 | 0.085 | 23.89 | 0.092 | 27.41 | 0.094 |
| | RATZ | 13.41 | 0.059 | 25.20 | 0.096 | 27.64 | 0.097 | 32.77 | 0.099 |
| | SPLICE | 10.02 | 0.044 | 18.47 | 0.054 | 19.11 | 0.087 | 21.14 | 0.092 |
| | MMCN | 9.26 | 0.043 | 19.24 | 0.063 | 20.14 | 0.086 | 23.08 | 0.091 |
| | SSM | 7.10 | 0.037 | 9.47 | 0.057 | 10.98 | 0.065 | 11.11 | 0.066 |
| | TRAJMAP | 7.09 | 0.036 | 8.04 | 0.045 | 9.08 | 0.046 | 10.48 | 0.052 |

**Table 4.2** Relative equal error rates of the proposed SV systems developed using the features derived from SFC methods

| Feature compensation methods | Relative equal error rate ($EER_R$) % | | | | | | | |
|---|---|---|---|---|---|---|---|---|
| | 0 dB SNR | | | | 5 dB SNR | | | |
| | Car | Factory | Pink | White | Car | Factory | Pink | White |
| RATZ | 30.32 | 5.35 | 4.66 | 3.58 | 35.99 | 7.18 | 9.05 | 4.07 |
| SPLICE | 47.69 | 26.67 | 31.90 | 29.98 | 52.17 | 31.97 | 37.11 | 38.11 |
| MMCN | 51.63 | 19.09 | 27.33 | 25.91 | 55.80 | 29.13 | 33.73 | 32.44 |
| SSM | 63.07 | 70.05 | 68.55 | 68.76 | 66.11 | 65.12 | 63.87 | 67.48 |
| TRAJMAP | 63.74 | 77.73 | 72.30 | 71.63 | 66.16 | 70.39 | 70.12 | 69.32 |

to perform moderately better than the SSM algorithm. The MinDCF values which varied in the range of 0.099–0.085 are reduced to the range 0.045–0.078. Compared to SSM, an EER drop as high as 1.9% is noticed in case of pink noise at 5 dB SNR, while the other cases closely follow by with reductions of 1.43% for factory noise at 5 dB, 1.31% for pink noise at 0 dB and 1.12% for white noise at 0 dB, respectively. The EER variance for TRAJMAP from 0 to 5 dB SNRs is much lower than the rest of the compared cases. This is an indication of the suitability of its application for SV tasks which are robust to SNR changes.

In order to demonstrate the performance improvement of the feature compensated SV framework over the baseline SV system in terms of EER, the 'Relative Equal Error Rate' ($EER_R$) given by $EER_R = \frac{(EER_B - EER_V)}{EER_B} \times 100\%$ is calculated where $EER_B$ and $EER_V$ are the equal error rates for the baseline and proposed SV systems, respectively. Table 4.2 shows the relative EER values of the proposed SV systems for different background environments.

The overall performance improvement gained by the use of feature compensation algorithms is apparent. An average relative EER of 12.52 % for RATZ, 37.07 % for SPLICE, 34.39 % for MMCN, 66.68 % for SSM and 69.67 % for TRAJMAP across all noisy environments is obtained.

## 4.5  Summary

In this chapter we demonstrated the significance of stochastic feature compensation methods for robust speaker verification in noisy environment. The effectiveness of the these data-driven methods was demonstrated for speaker verification in different simulated noisy environments. Recent state-of-the-art algorithms based on joint GMM modeling of clean and noisy data (i.e., SSM, TRAJMAP) were found to outperform well known algorithms like SPLICE and MMCN in terms of EER and minDCF metrics of speaker verification. The overall best performance was observed in case of the TRAJMAP method, which thereby suggests significance of dynamic feature correlation and robustness of long-term utterances towards background noise. Synthetic noisy data and clean utterances were used instead of actual stereo data in all experiments. For a better evaluation of the proposed method, actual stereo data may be used in future work. Data from real life environments at various other SNRs may be used instead of artificially constructed noisy data for a better insight into the efficiency of the proposed method.

# References

1. T. Kinnunen, Spectral features for automatic text-independent speaker recognition. PhD thesis, Department of Computer Science, University of Joensuu, 2004
2. D.A. Reynolds, Experimental evaluation of features for robust speaker identification. IEEE Trans. Speech Audio Process. **2**(4), 639–643 (1994)
3. R. Mammone, X. Zhang, R. Ramachandran, Robust speaker recognition: a feature-based approach. IEEE Signal Process. Mag. **13**(5), 58–71 (1996)
4. S. Furui, Cepstral analysis technique for automatic speaker verification. IEEE Trans. Acoust. Speech Signal Process. **29**(2), 254–272 (1981)
5. H. Hermansky, N. Morgan, RASTA processing of speech. IEEE Trans. Speech Audio Process. **2**(4), 578–589 (1994)
6. S. Boll, Suppression of acoustic noise in speech using spectral subtraction. IEEE Trans. Acoust. Speech Signal Process. **27**(2), 113–120 (1979)
7. A. Acero, R.M. Stern, Environmental robustness in automatic speech recognition, in *Proceedings of IEEE International Conference on Acoustics, Speech and Signal Processing (ICASSP '90)*, Albuquerque, 1990, vol. 2, pp. 849–852
8. S. Sarkar, K.S. Rao, Stochastic feature compensation methods for speaker verification in noisy environments. Appl. Soft Comput. **19**, 198–214 (2014). Elsevier
9. P.J. Moreno, B. Raj, R.M. Stern, Data-driven environmental compensation for speech recognition: a unified approach. Speech Commun. **24**(4), 267–285 (1998)
10. L. Deng, A. Acero, L. Jiang, J. Droppo, X. Huang, High-performance robust speech recognition using stereo training data, in *Proceedings of IEEE International Conference on Acoustics, Speech and Signal Processing*, Salt Lake City, 2001, vol. 1, pp. 301–304
11. M.J.F. Gales, P.C. Woodland, Mean and variance adaptation within the MLLR framework. Comput. Speech Lang. **10**, 249–264 (1996)
12. L. Buera, E. Lleida, A. Miguel, A. Ortega, Multi-environment models based linear normalization for speech recognition in car conditions, in *Proceedings of IEEE International Conference on Acoustics, Speech and Signal Processing (ICASSP '04)*, Montreal, 2004
13. M. Afify, X. Cui, Y. Gao, Stereo-based stochastic mapping for robust speech recognition. IEEE Trans. Audio Speech Lang. Process. **17**(7), 1325–1334 (2009)
14. C.M. Bishop, *Pattern Recognition and Machine Learning* (Springer, New York, 2006)
15. V. Digalakis, D. Rtischev, L. Neumeyer, E. Sa, Speaker adaptation using constrained estimation of Gaussian mixtures. IEEE Trans. Speech Audio Process. **3**(5), 357–366 (1995)
16. Y. Stylianou, O. Cappe, E. Moulines, Continuous probabilistic transform for voice conversion. IEEE Trans. Speech Audio Process. **6**(2), 131–142 (1998)
17. T. Toda, A.W. Black, K. Tokuda, Voice conversion based on maximum-likelihood estimation of spectral parameter trajectory. IEEE Trans. Audio Speech Lang. Process. **15**(8), 2222–2235 (2007)
18. H. Zen, Y. Nankaku, K. Tokuda, Stereo-based stochastic noise compensation based on trajectory GMMs, in *Proceedings of IEEE International Conference on Acoustics, Speech and Signal Processing (ICASSP '09)*, Taipei, 2009

19. K. Tokuda, T. Masuko, T. Yamada, T. Kobayashi, S. Imai, An algorithm for speech parameter generation from continuous mixture HMMs with dynamic features, in *Proceedings of the European Conference of Speech Communication Technology (EUROSPEECH '95)*, Madrid, Sept 1995, pp. 757–760
20. NIST-speaker recognition evaluations (1995) http://www.itl.nist.gov/iad/mig/tests/spk/
21. H. Hirsch, D. Pearce, The Aurora experimental framework for the performance evaluation of speech recognition systems under noisy conditions, in *Proceedings of the International Conference of Spoken Language Processing (ICSLP '00)*, Beijing, 2000
22. D. Reynolds, T. Quatieri, R. Dunn, Speaker verification using adapted Gaussian mixture models. Digit. Signal Process. **10**(1), 19–41 (2000)
23. A. Varga, H.J. Steeneken, Assessment for automatic speech recognition: II. NOISEX-92: a database and an experiment to study the effect of additive noise on speech recognition systems. Speech Commun. **12**, 247–251 (1993)

# Chapter 5
# Robust Speaker Modeling for Speaker Verification in Noisy Environments

**Abstract** The present chapter explores robust speaker modeling methods for speaker verification in noisy environment. The focus is specifically laid on building hybrid classifiers based on the combination of generative and discriminative models (e.g., Gaussian Mixture Models (GMMs) and Support Vector Machines (SVMs)). For improving the performance of the proposed speaker verification systems, utterance partitioning methods are used. The discussion is closely followed by state-of-the-art variants of GMM supervector based approaches (i.e., i-vectors) and algorithms for combining robust classifiers.

The application of stochastic feature compensation for speaker verification (SV) as studied in Chap. 4, is associated with certain drawbacks. Firstly, it depends on the availability of stereo data which is expensive to acquire. Secondly, a priori knowledge about a speaker's test environment is assumed i.e., the background environment during evaluation should be reflected in the stereo training data. Lastly, substantial amount of data may be required for the joint probability modeling techniques. However, in real-life scenarios the test environments are often unknown and time-varying (non-stationary). SV applications deployed in hand-held devices are additionally expected to perform in real-time with minimal data requirements.

As an alternative strategy, model compensation and robust speaker modeling methods can be explored. The role of these two methods have been briefly explained in Chaps. 1 and 2, respectively. We had also emphasized on certain limitations of the conventional model compensation methods such as requirement of clean speaker models, dependence on a mathematical representation of the noise corruption process. Additionally, popular model compensation methods like Parallel Model Combination demand substantial amount of training data and high computational resources which may not be frequently available.

The present chapter explores robust speaker modeling methods for SV in noisy environment. The focus is specifically laid on building hybrid classifiers based on the combination of generative and discriminative models (e.g., Gaussian Mixture

K.S. Rao and S. Sarkar, *Robust Speaker Recognition in Noisy Environments*,
SpringerBriefs in Electrical and Computer Engineering,
DOI 10.1007/978-3-319-07130-5_5, © The Author(s) 2014

Models (GMMs) and Support Vector Machines (SVMs)). The discussion is closely followed by state-of-the-art variants of GMM supervector based approaches (i.e., i-vectors) and algorithms for combining robust classifiers.

## 5.1  GMM-SVM Combined Approach for Speaker Verification

The traditional GMM-UBM based speaker verification system requires a Universal Background Model (UBM) [1] and a *Maximum aPosteriori* (MAP) adapted Gaussian Mixture Model (GMM) to represent the impostor and actual speaker classes, respectively. During the evaluation stage, a test utterance is classified based on its statistical similarities with the claimed target speaker model (GMM) and the background model (UBM). Gaussian Mixture Models (GMMs) are extensively applied for speaker modeling due to their strong probabilistic framework, scalability to large training sets and high recognition accuracy. GMMs belong to the family of generative models in which each speaker is modeled individually. Performance accuracy of a SV system is usually increased when these generative models are brought into a discriminative framework using Support Vector Machines (SVMs) [2].

SVMs have been established as an effective discriminative classifiers for speaker recognition tasks [3]. Through a non-linear function (i.e., kernel) a SVM maps input vectors to a high dimensional space where classes are more likely to be linearly separable [4]. However, fixed length representation of utterance is crucial for SVM training in order to avoid large target models and slow scoring. This had initially led to concept of 'sequence kernels'[5] where variable length utterances were mapped to fixed length vectors. A robust representation was proposed later in which fixed size 'supervectors' constructed by stacking the means of MAP adapted GMMs were used as an input to SVM kernels [2]. Conventionally, a GMM based system calculates log-likelihood probabilities (scores) of features extracted at a frame level. In contrast, supervectors provide numerical comparison of speech utterances as an entire sequence rather than frame-wise probabilities thus preserving information which can be otherwise discarded in the frame-level [5]. Supervectors are attractive due to a number of reasons. Besides providing a high-dimensional representation for SVM classification, supervectors can distinctly characterize speaker and channel information [6]. Additionally, they can be used to compensate for channel and session variabilities [7]. In this chapter we shall explore the robustness of supervector based speaker modeling approaches for SV in noisy environment. In the following sections we briefly introduce SVMs and describe the process of integrating GMM supervectors in the SVM framework. Figure 5.1 shows the various stages of a GMM-SVM based SV system, each of which are elaborated in the remaining part of the section.

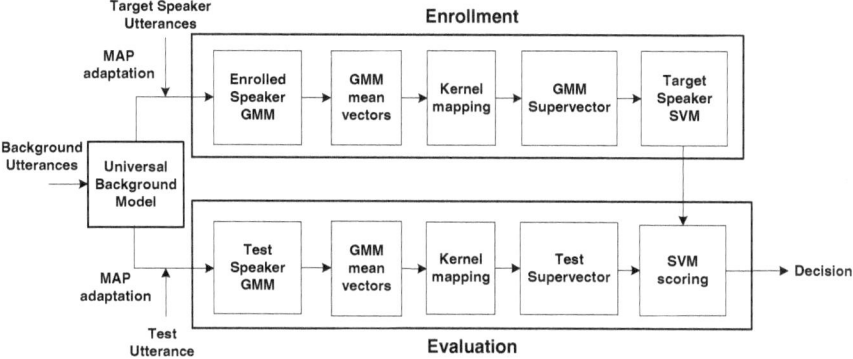

**Fig. 5.1**   Block diagram of the GMM-SVM framework for speaker verification

### 5.1.1   Support Vector Machines

A support vector machine (SVM) is a binary classifier [4]. Using labeled training vectors, a SVM optimizer finds a decision hyperplane that maximizes the margin of separation between two classes (target speaker and impostor). The classifier equation is given as follows:

$$y(x) = \sum_{i=1}^{L} \alpha_i t_i K(x_i, x) + d \qquad (5.1)$$

where $\alpha_i > 0$ are the Lagrange multipliers and $x_i$ are the Support vectors. Both these parameters are learned during the optimization process. $t_i \in \{-1, +1\}$ are the training labels, $K$ is the desired kernel mapping and $d$ is a bias parameter. For any input vector $x$ the actual output $y(x)$ is compared with a decision threshold for final classification. The kernel function is constrained to satisfy the Mercer's conditions [4], so that they can be expressed as

$$K(x, y) = S(x)^T S(y) \qquad (5.2)$$

where $S(x)$, $S(y)$ are high dimensional mappings for inputs $x$ and $y$, respectively.

### 5.1.2   Construction of GMM Supervectors

The GMM-UBM framework for SV was discussed in Chap. 3. During enrollment, the pre-estimated parameters of the UBM (i.e., mean and covariance (optionally)) are modified by MAP adaptation using a target speaker's utterance to produce speaker specific GMMs given by the following equation

$$p(x) = \sum_{i=1}^{M} w_i \mathcal{N}(x; m_i, \Sigma_i) \qquad (5.3)$$

where $m_i$, $\Sigma_i$ denotes the mean and covariance of the $i$th multivariate Gaussian component $\mathcal{N}()$ and $M$ is the total number of GMM components. The high-dimensional vector obtained by concatenating the mean vectors $m_i$ of each Gaussian is generally termed as a *supervector*. Therefore $D$ dimensional feature vectors in the input space are converted to a single $M \times D$ dimensional supervector irrespective of the number of feature vectors available. In other words, this process transforms variable length utterances to a unique fixed-size vector which carries speaker information. This representation is in conformity with Eq. (5.2) where two arbitrary utterances $a$ and $b$ from the input space can be compared in the supervector space using the relation $K(a, b) = S(a)^T S(b)$, where $K$ is the kernel function and $S(a)$, $S(b)$ are the supervectors obtained from utterances $a$ and $b$, respectively. The supervector construction process can be summarized as follows [8]

1. A target speaker GMM is obtained by MAP adaptation of the UBM using the speaker's enrollment utterance.
2. A kernel function is used to transform parameters of each GMM component to a fixed length vector. The vector corresponding to the $i$th GMM component constitutes the $i$th subvector of a supervector.
3. All the subvectors are concatenated to obtain a high-dimensional supervector.

### 5.1.3  SVM Kernels

The main design component in an SVM is the kernel, which is an inner product in the SVM feature space. The basic goal in SVM kernel design is to find an appropriate metric in the SVM feature space relevant to the classification problem. In this section we define the kernels used in our work.

#### 5.1.3.1  KL Divergence Kernel

The Kullback Leibler divergence (KL div) is a non-symmetric distance measure between two probability distributions. Given two distributions $p_a$ and $p_b$, the KL divergence between them is defined as

$$D_{KL}(p_a, p_b) = \int p_a(x) \log \left( \frac{p_a(x)}{p_b(x)} \right) dx \qquad (5.4)$$

However the KL divergence doesn't satisfy the Mercer's condition for a valid kernel. As a solution a symmetrized version of the KL divergence, obtained by bounding the expression by log-sum inequality was proposed as a kernel in [2]. The final version was a linear function of two MAP adapted GMMs $p_a$ and $p_b$ corresponding to utterances $a$ and $b$. Ignoring adaptation of the UBM covariance matrix $\Sigma_i^u$ and weights $w_i$, the resulting Kernel is given by

$$K_{KL}(p_a, p_b) = \sum_{i=1}^{M} (\sqrt{w_i}(\Sigma_i^u)^{-1/2}m_i^a)^T (\sqrt{w_i}(\Sigma_i^u)^{-1/2}m_i^b) \qquad (5.5)$$

where $m_i^a$ and $m_i^b$ are the $i$th component means of $p_a$ and $p_b$ respectively. Thus the $i$th subvector of the GMM supervector for any utterance $\lambda$ is given by

$$s_i^\lambda = \sqrt{w_i}(\Sigma_i^u)^{-1/2}m_i^\lambda \quad i = 1, 2, \ldots, M$$

The final supervector obtained by concatenating the subvectors is given by $S^\lambda = [s_1^T, s_2^T, \ldots s_M^T]^T$.

### 5.1.3.2   GMM-UBM Mean Interval Kernel

The Bhattacharya distance between two probability distribution $p_a$ and $p_b$ is given by

$$D_{Bhatt}(p_a, p_b) = \int \sqrt{p_a(x)p_b(x)}dx \qquad (5.6)$$

For multivariate Gaussian distributions, computing this measure requires estimation of the covariance matrices which in turn demands a high amount of training data. Hence this measure is avoided in practical scenarios. However, it was shown in [9] that second order statistics derived from limited amount of training data could provide supplementary discriminative information, when used effectively. The GMM-UBM Mean Interval (GUMI) kernel based on the Bhattacharya distance between GMMs $p_a$ and $p_b$, as proposed in [9] is given by

$$K_{GUMI}(p_a, p_b) = \sum_{i=1}^{M} (m_i^b - m_i^a)^T \left[ \frac{\Sigma_i^a + \Sigma_i^b}{2} \right]^{-1} (m_i^b - m_i^a) \qquad (5.7)$$

Considering the statistical similarities of a adapted speaker GMM and the UBM the $i$th subvector of the GMM supervector for an utterance $\lambda$ is given by

$$s_i^\lambda = \left[ \frac{\Sigma_i^\lambda + \Sigma_i^u}{2} \right]^{-1/2} (m_i^\lambda - m_i^u) \quad i = 1, 2, \ldots, M$$

The final supervector obtained by concatenating the subvectors is given by $S^\lambda = [s_1^T, s_2^T, \ldots s_M^T]^T$.

## 5.1.4  SVM Scoring

Given a supervector $S^{test}$ derived from a test utterance $X^{test}$ the Kernel scoring is obtained as follows:

$$Score(X^{test}) = \sum_{i=1}^{L} \alpha_i t_i K(X^i, X^{test}) + d = \left( \sum_{i=1}^{L} \alpha_i t_i S^i \right)^T S^{test} + d \quad (5.8)$$

where $X^i$ are the sequence of learned support vectors, $S^i$ are the supervectors corresponding to $X^i$, $\alpha_i$ are the non-zero Lagrange multipliers and $t_i \in \{-1, +1\}$ depending on the class of vector $X^i$. $L$ is the total number of support vectors and $d$ is a bias term. $K$ is either of the two kernels used and $T$ denotes matrix transpose.

## 5.1.5  Experimental Setup

All experiments are conducted on the NIST-SRE-2003 database. The data consists of single training utterances of approximately 2 min length from each of 356 enrolled speakers and 3,500 test utterances (approximately 10–15 s each) for evaluation. The stages involved in developing the GMM-SVM based SV system are briefly discussed in the following sections.

### 5.1.5.1  Background Simulation and Feature Extraction

The background simulation and feature extraction process has already been discussed in Chaps. 3 and 4. Summarily, all training and test utterance were degraded with additive noises (car, factory, pink and white) collected from the NOISEX-92 database. Two types of background simulations were carried out viz., (i) uniform backgrounds in which an entire utterance (training/testing) was degraded with a particular type of noise at 0, 5 and 10 dB SNRs and (ii) varying backgrounds in which non-overlapping segments of an utterance (training/testing) were individually degraded with a specific type of noise at 0, 5, 7 and 10 dB SNRs. After an energy-based voiced activity detection, 39-dimensional feature vectors (consisting of 13 MFCCs + $\Delta$ + $\Delta\Delta$ excluding $C_0$) derived from a 26 channel mel-scaled filterbank, were extracted from pre-emphasized speech frames of 20 ms with a frame-overlap of 10 ms. All feature vectors were subjected to cepstral mean subtraction followed by cepstral variance normalization.

#### 5.1.5.2   Speaker Modeling

A 1,024-component GMM constructed from 20 h of speech (10 h male + 10 h female) collected from the SwitchBoard II corpus, was used as the UBM. Three hundred and fifty-six target speaker GMMs were obtained by MAP-adaptation of the UBM using the noise-degraded enrollment utterances in each dataset described in Sect. 5.1.5.1. A GMM supervector was constructed from each target speaker GMM as described in Sect. 5.1.2. The kernels described in Sect. 5.1.3 were individually used for mapping. The supervectors obtained were of 39,936 dimension (1,024 mixtures × 39 dim mean). For discriminative modeling each target speaker in a dataset was distinguished from the remaining 355 background speakers (impostors). A SVM for each speaker was trained with the speaker's supervector labelled as +1 and the background supervectors labelled as −1, respectively. The KL divergence and GMM-UBM mean interval kernels were used for SVM training as described in Sect. 5.1.1.

### 5.1.6   Performance Evaluation

All experiments were performed in matched condition i.e., training and evaluation phases having similar backgrounds. An additional evaluation was performed in clean condition. The 3,500 test utterances in each noise-corrupted dataset were transformed to supervectors prior to SVM scoring (Eq. 5.8). The NIST-2003 primary task was carried out in which each noisy test utterance (supervector) was evaluated against 11 target speaker models (SVMs) from the same dataset. The equal error rate (EER) and minimum DCF (MinDCF) values were used as metrics for performance evaluation. The standard GMM-UBM based SV systems have been used as a baseline system for performance comparison.

Table 5.1 summarizes the performance of the various SV systems developed in uniform noisy environments. The improvement in performance accuracy is clearly apparent in case of the GMM-SVM based systems in comparison to the baseline. This is manifested by a consistent reduction in EER and MinDCF values across all 12 types of noisy environments. The performance accuracy is observed to degrade non-uniformly with decreasing SNR levels. The loss in accuracy of the GMM-SVM based systems with increasing noise distortion, is correlated with that of the baseline system. An average increment of 4.48, 3.69 and 3.93 % EER values is observed for a transition from 10 to 5 dB SNR in case of the baseline, GMM-SVM (KL div) and GMM-SVM (GUMI), respectively across all four backgrounds. The same observation sequence for the 5 to 0 dB SNR transition shows average increments of 2.12, 1.07 and 0.98 % EERs which indicates that the SVM based systems are relatively more robust towards noise degradations. However the averaged metric values does not characterize individual noise behavior. For instance we observe fractional performance improvement in case of factory noise at 0 and 5 dB for GMM-SVM (KL div). A general order of precedence (best to worst) of the noisy

**Table 5.1** Performance of the SV systems in uniform background environments and clean conditions

| SNR | Noises | GMM-UBM | | GMM-SVM (KL div) | | GMM-SVM (GUMI) | |
|---|---|---|---|---|---|---|---|
| | | EER (%) | MinDCF | EER (%) | MinDCF | EER (%) | MinDCF |
| 0 dB | Car | 18.04 | 0.071 | 14.32 | 0.063 | 13.82 | 0.058 |
| | Factory | 23.17 | 0.089 | 21.27 | 0.086 | 20.32 | 0.086 |
| | Pink | 26.65 | 0.097 | 21.23 | 0.087 | 19.92 | 0.085 |
| | White | 30.98 | 0.097 | 22.72 | 0.076 | 22.67 | 0.079 |
| 5 dB | Car | 18.11 | 0.071 | 13.77 | 0.059 | 13.10 | 0.058 |
| | Factory | 20.96 | 0.085 | 20.87 | 0.085 | 19.92 | 0.084 |
| | Pink | 23.89 | 0.092 | 19.65 | 0.086 | 19.06 | 0.085 |
| | White | 27.41 | 0.094 | 20.96 | 0.074 | 20.73 | 0.072 |
| 10 dB | Car | 15.44 | 0.068 | 12.24 | 0.047 | 11.38 | 0.046 |
| | Factory | 16.44 | 0.072 | 14.81 | 0.053 | 14.50 | 0.052 |
| | Pink | 18.65 | 0.081 | 15.67 | 0.058 | 15.72 | 0.057 |
| | White | 21.91 | 0.087 | 17.75 | 0.065 | 15.49 | 0.067 |
| Clean | | 06.93 | 0.033 | 06.72 | 0.030 | 06.44 | 0.030 |

backgrounds is noticed in terms of overall performance of the GMM-SVM based systems. Ignoring minor exceptions in case of 0 and 10 dB SNRs the order is car, pink, factory and white. This is in contrast with the baseline where the performance in pink noisy background is worse than that of factory background for all SNR levels. A comparison amongst the GMM-SVM based systems reveals that the SVMs with GUMI kernel performs moderately better than those with KL div kernel with an average reduction of 0.72 % EER across all environments.

Figure 5.2 shows the DET plots for the SV systems in (a) Car (b) Factory (c) Pink and (d) White noisy backgrounds at various SNRs. The DET curves of the GMM-UBM and GMM-SVM based systems are denoted by a set of black, red (GUMI) and blue (KL div) lines, respectively. The red and blue lines show a shift towards the origin indicating joint reduction of error probabilities. Additionally, a distinct anticlock-wise rotation in the red and blue set of curves can be noticed in comparison to the black curves (baseline) which is particularly prominent in case of factory, pink and white noise. This characteristic suggests higher reduction in 'miss' error rates compared to the 'false alarm' rates which is also evident from significant reduction in MinDCF values. Table 5.2 shows the performance improvement of the GMM-SVM based systems compared to the baseline in terms of the 'Relative Equal Error Rate' ($EER_R$) defined as $EER_R = \frac{EER_B - EER_V}{EER_B} \times 100\,\%$ where $EER_B$ and $EER_V$ are the EER values of the baseline (GMM-UBM) and GMM-SVM based systems, respectively. The SV systems with KL div kernels score average $EER_R$ values of 21.17, 6.18, 18.02 and 23.06 % for car, factory, pink and white noisy backgrounds, respectively. The GUMI kernel based SV systems perform even better with average $EER_R$ values of 25.78, 9.69, 20.39 and 26.83 % in the same backgrounds. Figure 5.3 shows the changes in (a) EER and (b) Relative EER of the SV systems at different SNRs in uniform noisy environments.

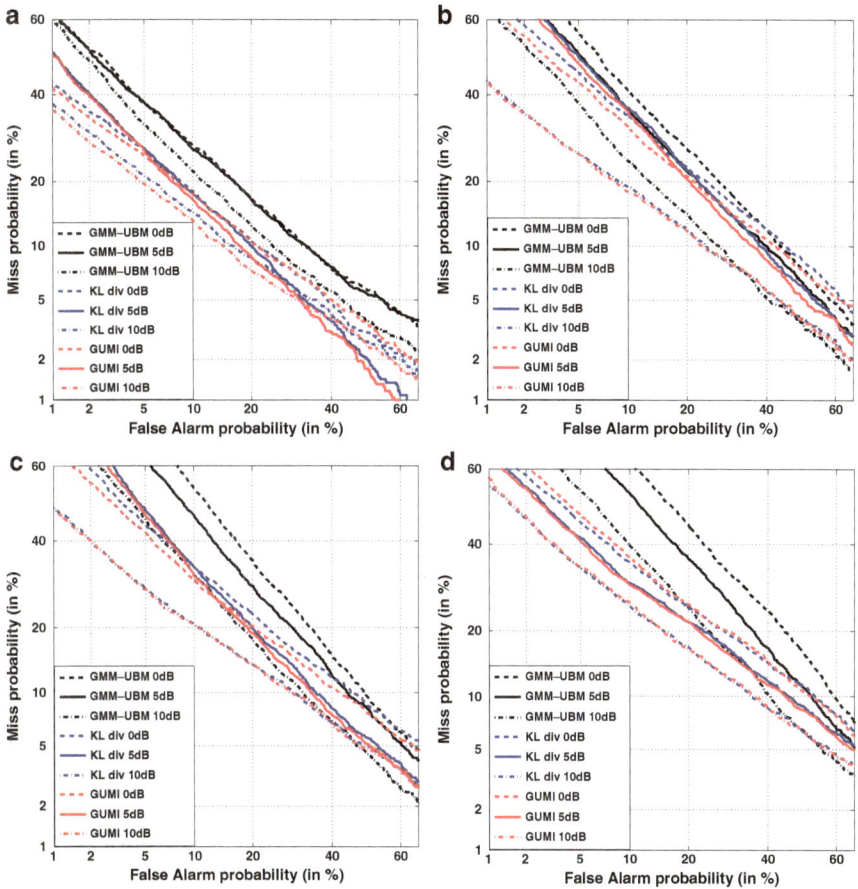

**Fig. 5.2** DET plots of the SV systems in uniform background environment with (**a**) car (**b**) factory (**c**) pink and (**d**) white noise. The *black*, *blue* and *red colors* indicate GMM-UBM, GMM-SVM trained using KL div kernel and GUMI kernel, respectively

The EER values reduce consistently with increasing SNRs. However, the individual $EER_R$ values across each SNR shows distinct behavior for each noise. In most cases there is an abrupt change at the 5 dB SNR level with the exception of pink noise which shows a consistent linear reduction for both types of SVMs.

Table 5.3 summarizes the performance of the SV systems developed in varying background environments. Though a direct comparison is inappropriate, an overall inferior performance is observed in contrast to SV systems in uniform noisy backgrounds. The utterances used for training these systems had short segments corrupted with the noises individually used for uniform background simulation, at a fixed SNR (see Chap. 3). Thus the average SV performance across all uniform backgrounds at a fixed SNR was compared with the SV performance in varying

**Table 5.2** Relative equal error rates for GMM-SVM based SV systems in uniform background environments

| | Relative equal error rate $EER_R$ (%) | | | | | |
| | SNR (0 dB) | | SNR (5 dB) | | SNR (10 dB) | |
| Noises | KL div | GUMI | KL div | GUMI | KL div | GUMI |
|---|---|---|---|---|---|---|
| Car | 20.62 | 23.39 | 23.96 | 27.66 | 20.73 | 26.30 |
| Factory | 08.20 | 12.30 | 00.43 | 04.96 | 09.91 | 11.80 |
| Pink | 20.38 | 25.25 | 17.75 | 20.22 | 15.98 | 15.71 |
| White | 26.66 | 26.82 | 23.53 | 27.41 | 18.99 | 29.30 |

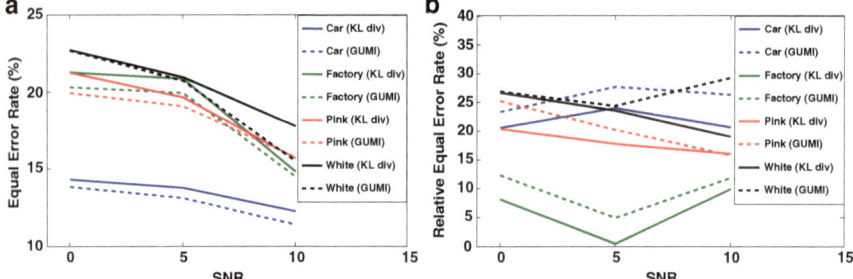

**Fig. 5.3** (**a**) Equal error rates and (**b**) relative equal error rates of GMM-SVM based SV systems at different SNRs in uniform background environments

**Table 5.3** Performance of the SV systems in varying background environments

| SNR (dB) | GMM-UBM | | GMM-SVM (KL div) | | GMM-SVM (GUMI) | |
| | EER (%) | MinDCF | EER (%) | MinDCF | EER (%) | MinDCF |
|---|---|---|---|---|---|---|
| 0 | 27.05 | 0.094 | 23.48 | 0.086 | 22.76 | 0.085 |
| 5 | 25.74 | 0.086 | 22.18 | 0.080 | 21.32 | 0.081 |
| 7 | 25.29 | 0.083 | 19.74 | 0.073 | 19.11 | 0.071 |
| 10 | 21.86 | 0.080 | 18.65 | 0.072 | 16.44 | 0.069 |

background at the same SNR. The average EER values of the baseline systems across uniform backgrounds obtained earlier (see Table 5.1) were 24.70, 22.59 and 18.11 % for 0, 5 and 10 dB SNRs, respectively. Similarly, average EER values for the GMM-SVM (KL div) and GMM-SVM (GUMI) systems at the three SNR levels were 19.89, 18.81, 15.12 % and 19.18, 18.20, 14.28 %, respectively. In contrast, performance of the baseline systems in varying backgrounds shows an average EER increment of 3.08 %, ignoring the 7 dB SNR value. A likewise comparison with the corresponding SVM based systems with KL div and GUMI kernels, shows average increments of 3.50 and 2.95 % EERs, respectively. A possible explanation to this behavior is the inadequate amount of data used for capturing the statistics of non-stationary noise. The rapid change in noise could also causes a greater

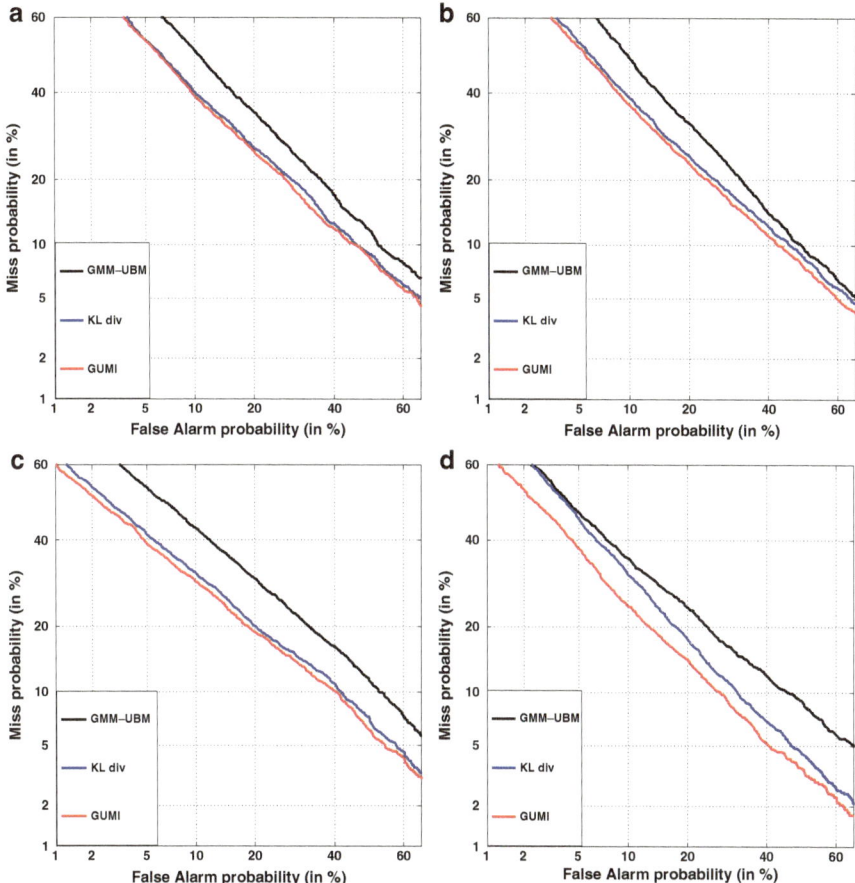

**Fig. 5.4**   DET plots for the SV systems in varying background environments at (**a**) 0 dB (**b**) 5 dB (**c**) 7 dB and (**d**) 10 dB SNRs. The *black*, *blue* and *red colors* indicate GMM-UBM, GMM-SVM trained using KL div kernel and GUMI kernel, respectively

degree of mismatch during the evaluation phase. The effect is more prominent in case of the baseline systems and comparatively less for the others. However, certain changes in the behavior of the GMM-SVM based systems are apparent. Unlike the uniform background case, the use of costly covariance kernels (GUMI) provides a better improvement of 1.11 % EER over the KL div kernels, when averaged across all SNRs.

Figure 5.4 shows the DET plots of the SV systems developed in varying background environment at (a) 0 dB (b) 5 dB (c) 7 dB and (d) 10 dB SNRs. The characteristics of the blue (KL div) and red (GUMI) curves are in contrast to those in Fig. 5.2. In most cases, there are no apparent rotation in the curves though an overall shift towards the origin can be noticed. In fact, the red and blue lines shows a slight rotation in clock-wise direction for 10 dB SNR despite preserving a notable difference in false alarm rates with respect to the baseline. Interestingly, the set of red and blue lines show similar properties in terms of the slope, shape and alignment with each other. The overall improvement in average MinDCF values are $8 \times 10^{-3}$ and $9.25 \times 10^{-3}$ for the SVM based systems with KL div and GUMI kernels, respectively. This is significantly lower in comparison to the uniform background scenarios. The overall inferior performance of the SV systems in varying background environment encouraged the use of a SVM-based channel compensation method prior to SVM training.

### 5.1.6.1   Nuisance Attribute Projection

Nuisance Attribute Projection (NAP) [10] is a commonly applied session compensation technique for GMM-SVM based SV systems. NAP aims to remove components (nuisance attributes) from the supervector space which are irrelevant for speaker recognition and may carry information related to channel, background etc. In other words, it eliminates the subspace which causes variabilities. This is achieved by an orthogonal projection of the supervectors in the channel's complementary space. A projection matrix $P$ is trained using an auxillary set of speakers carrying various channel information as given by

$$P = I - vv^T \tag{5.9}$$

where $v$ is a low rank rectangular matrix whose columns are given by 'k' eigenvectors with highest eigenvalues of the supervector's within-class covariance matrix. Thus a new linear kernel is constructed for inputs $x$ and $y$ after NAP operation on the supervectors $S(x)$ and $S(y)$ which is given by

$$K(x, y) = [PS(x)]^T [PS(y)] \tag{5.10}$$

The formal steps of calculating the NAP projection matrix are given as follows

1. A set of supervectors is constructed from a target speaker's enrollment utterances.
2. For each speaker, the mean of the supervectors is subtracted from each supervector in the set to subdue intra-speaker variability.
3. A large matrix $V$ is formed whose columns constitute mean-removed supervectors from all speakers. This matrix is expected to contain session information.
4. The within class covariance matrix $W$ of matrix $V$ is calculated as $W = VV^T$ and subjected to eigen decomposition.
5. The eigenvectors having the largest 'k' eigenvalues are used to form the rectangular matrix $v$. The integer 'k' (called NAP rank), is usually determined empirically.
6. The projection matrix $P$ is calculated by Eq. (5.9).

**Table 5.4** Comparison of the performances of the GMM-SVM based SV systems in varying background environments with and without NAP compensation

| SNR (dB) | GMM-SVM | | | | GMM-SVM + NAP | | | |
|---|---|---|---|---|---|---|---|---|
| | KL div | | GUMI | | KL div | | GUMI | |
| | EER (%) | MinDCF | EER (%) | MinDCF | EER (%) | MinDCF | EER (%) | MinDCF |
| 0 | 23.48 | 0.086 | 22.76 | 0.085 | 22.22 | 0.084 | 21.27 | 0.083 |
| 5 | 22.18 | 0.080 | 21.32 | 0.081 | 21.05 | 0.079 | 20.14 | 0.080 |
| 7 | 19.74 | 0.073 | 19.11 | 0.071 | 18.74 | 0.072 | 18.06 | 0.070 |
| 10 | 18.65 | 0.072 | 16.44 | 0.069 | 17.75 | 0.070 | 15.77 | 0.067 |

The NAP matrix was trained using 400 utterances collected from a set of 100 speakers of the NIST-SRE-2004 corpus. Steps 1–6 define the ideal method for estimating the NAP matrix. However, a direct application of Step 4 was infeasible due to the large size of supervectors (i.e., 39,936). As an alternative strategy, an eigenvector matrix $v'$ was first constructed by eigen decomposition of the matrix $W' = \frac{1}{N} V^T V$ where $N$ is the number of supervectors. The required matrix $v$ was then obtained by the operation $v = N^{-1/2} V v' \lambda^{-1/2}$ where $\lambda$ is a diagonal matrix containing eigenvalues of the matrix $W'$. NAP transformation produced four new sets of supervectors (one for each SNR), which were subjected to SVM training and evaluation as explained earlier in Sects. 5.1.5 and 5.1.6, respectively. A NAP rank of 80 was empirically chosen to produce best results. Table 5.4 summarizes the performance of the GMM-SVM based SV systems after NAP compensation. Marginal improvements in EER and MinDCF values are noticed, in comparison to the initial set of observations. The average EER reduction for the new set of SVMs in comparison to their earlier version (columns 2 and 4 of Table 5.4) are 1.07 % (Kl div) and 1.10 % (GUMI), respectively. The EER improvements in comparison to the baseline are 5.05 and 6.18 % for KL div and GUMI kernels, respectively. The improvements due to NAP are observed to diminish consistently with increasing SNR. This can be easily interpreted from the sequence of EER reductions for the KL div based systems given by 1.26, 1.13, 1.00 and 0.90 % for 0, 5, 7 and 10 dB SNRs, respectively. The same sequence for the GUMI based systems is 1.49, 1.18, 1.05 and 0.67 %.

Figure 5.5 shows the effect of NAP in the DET curves of the GMM-SVM based SV systems. The broken blue and red lines has been used to denote the NAP based GMM-SVM systems with KL div and GUMI kernels, respectively. No significant changes are apparent in the broken lines except for the consistent shift towards the origin which results in appropriate reduction in MinDCF values. In most cases, the KL div systems with NAP performs better than the GUMI based systems without NAP with an exception in the 10 dB SNR case. The overall improvement in average MinDCF values are $9.25 \times 10^{-3}$ and $9.75 \times 10^{-3}$ for SVM based systems with KL div and GUMI kernels, respectively.

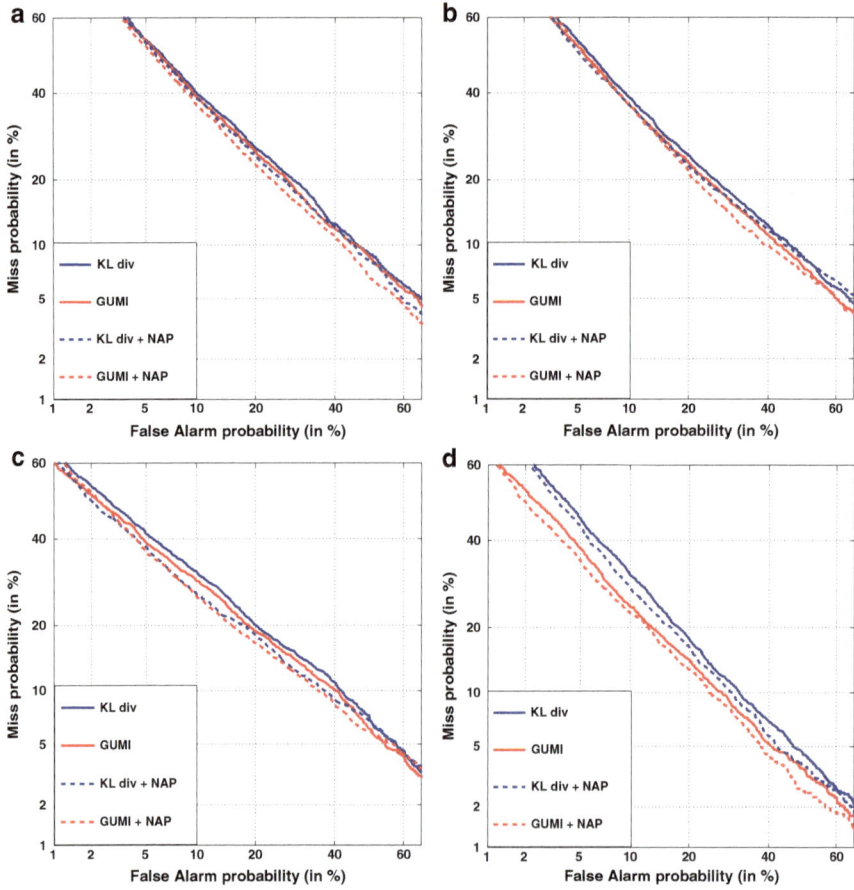

**Fig. 5.5** DET plots showing the effect of NAP in the GMM-SVM based SV systems in varying background environment at (**a**) 0 dB (**b**) 5 dB (**c**) 7 dB and (**d**) 10 dB SNRs

Table 5.5 summarizes the relative EER values of the GMM-SVM based SV systems developed without and with NAP. An average relative EER of 20.19 and 24.89 % is obtained for the SVM based systems with KL div and GUMI kernels, respectively. However the improvement due to NAP is substantially limited with average relative EER increments of only 4.28 and 4.32 %, respectively for the aforementioned systems. The characteristics of (a) EERs and (b) Relative EERs of the GMM-SVM based systems at various SNRs in varying background environments has been shown in Fig. 5.6.

| | Relative equal error rate ($EER_R$) (%) | | | |
|---|---|---|---|---|
| | GMM-SVM | | GMM-SVM + NAP | |
| SNR (dB) | KL div | GUMI | KL div | GUMI |
| 0 | 13.20 | 15.86 | 17.86 | 21.37 |
| 5 | 13.83 | 17.17 | 18.22 | 21.76 |
| 7 | 21.95 | 24.44 | 25.90 | 28.59 |
| 10 | 14.68 | 24.79 | 18.80 | 27.86 |

**Table 5.5** Relative equal error rates for GMM-SVM based SV systems in varying background environments

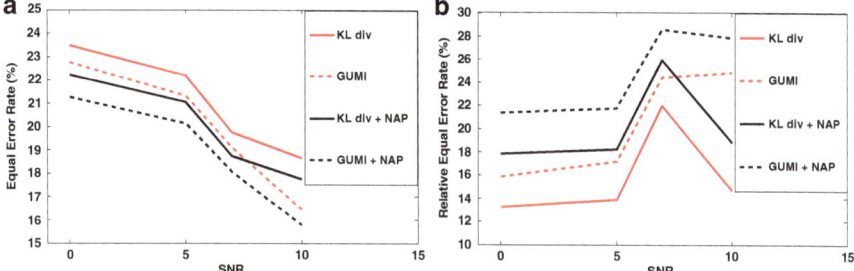

**Fig. 5.6** (**a**) Equal error rates and (**b**) relative equal error rates of the GMM-SVM based SV systems at different SNRs in varying background environments

## 5.2  Utterance Partitioning for Improving GMM-SVM Based Speaker Verification Performance

The studies conducted in various types of noisy environments, as described in the previous section, unanimously indicates that the SV performance accuracy enhances with the use of GMM supervectors in conjunction with SVMs. However, it was also noticed that the performance improvements were not consistent across different noisy backgrounds at various SNR levels. In fact, fractional changes in EER values were observed in quite a number of cases e.g., uniform factory noise at 5 dB SNR. Besides, the use of GUMI kernels yielded marginal improvements in comparison to the standard KL div kernels, in most of the simulated environments. Contrary to expectations, the benefits of the complex NAP operations were also nominal. These phenomena suggested scope for further improvement in the standard SV system design. Instead of exploring alternative modeling methods, a number of inherent drawbacks in the existing method were addressed for a change. A few of such drawbacks can be highlighted as follows

- **Data imbalance:** A distinct aspect of the conventional SVM training method is that the number of background utterances (supervectors) vastly outnumber the number of enrolment utterances from a target speaker (typically one). This obviously leads to the generation of a larger number of support vectors in the majority class (background speakers) compared to the minority class (target speaker) causing a phenomenon called 'data imbalance' [11, 12]. As a consequence, the SVM decision boundary skews towards the minority class which causes high false rejection ('miss') rates during the kernel scoring in the evaluation phase (Eq. 5.8) of SV, unless the decision threshold is properly adjusted to compensate for the bias.

- **Mismatched utterance lengths:** The duration of training and test utterances plays a significant role in SV accuracy [13]. The amount of available training data in utterances determines the degree of MAP-adaptation of a GMM and thus affects the composition of the supervectors as discussed in Sect. 5.1. The difference in the enrolment and test utterance lengths (the former being considerably larger than the latter), can thus lead to statistical mismatches during the evaluation phase. In fact, prior studies have shown the benefits of matching training and test utterance durations for SV [14]. Additionally, recent studies have revealed that the discriminative power of fixed-size vectors used for representing variable length utterances saturates when the utterance length exceeds a threshold (typically 2 min) [15]. In such situations, the excess data can be utilized by generating new vectors rather than a single one.

- **Small sample-size problem:** In a typical training dataset, the number of speakers could be fairly large, but the number of available sessions per speaker are often quite limited. When the number of training speakers or the number of recording sessions per speaker are insufficient, numerical errors occur in estimating transformation matrices associated with the construction of supervectors (e.g., NAP), resulting in inferior performance (as noticed in Sect. 5.1.6). In machine learning literature, this is known as the 'small sample-size problem' [16, 17].

The various available strategies used to mitigate the effect of 'data imbalance' can be broadly categorized as (i) data processing approaches and (ii) algorithmic approaches. The family of methods in the first group tries to reduce the disproportionate ratio of support vectors in each class [18]. This can be done by (a) Over-sampling methods, where new training examples are generated from the existing minority class data [19, 20] (b) Under-sampling methods, where a subset of majority class examples are used to train individual SVMs [21, 22] and (c) a combination of Over-sampling and Under-sampling [12]. Under-sampling is usually not preferred for SV tasks since it causes loss of discriminative information whereas over-sampling methods are a trade-off between improved classification accuracy and increased computational load. The algorithmic approaches modify the classifier algorithm to counter data imbalance. Earlier methods assigned asymmetric misclassification costs to the positive and negative training examples [23] which was marginally effective since the Lagrange multipliers in both classes were scaled to satisfy a SVM constraint. Other methods modified the kernel function according to the data distribution which lead to complex training procedures [24].

The mismatch in utterance durations as highlighted earlier can be resolved by either using shorter length training/enrollment utterances or longer test utterances. In the former case, the major issue is to empirically determine an appropriate length of training utterances which can contribute towards MAP adaptation without sacrificing representative power. Lengthy test utterances as an alternative are usually not preferred for real-life applications. Handling the small sample-size problem is also subject to practical constraints such as availability of a co-operative set of speakers for multi-session recordings or requesting multiple enrolment utterances from client speakers etc.

As a solution to the aforementioned problems a synthetic data generation technique using partitioned utterances as proposed in [25], was applied in the present work. Specifically, the sequence of frames in an utterance were randomized followed by dividing it into a number of fixed-length sub-utterances which were individually used for supervector construction. The formal steps of the method, known as Utterance Partitioning with Acoustic Vector Resampling (UP-AVR), are briefly outlined as follows:

1. Given an enrollment utterance of a target speaker, the acoustic vectors (MFCCs) are computed and their sequence of occurrence (frame indices) in the utterance are randomized. This randomized sequence is then divided into $N$ partitions (sub-utterances).
2. Steps 1 is repeated $R$ times to produce $RN$ sub-utterances.
3. Each of the sub-utterances produced in Step 2 together with the original utterance are individually used for supervector construction. Thus a total of $RN + 1$ target speaker supervectors are obtained.
4. Each background utterance is like-wise partitioned into $N$ sub-utterances as given in Step 1. However, unlike the enrollment utterances, Step 2 is skipped and Step 3 is directly applied instead.
5. For $B$ background utterances, a total of $B(N + 1)$ background supervectors are thus obtained.

Based on the length of available utterances in the present work, parameter values of $N = 2$ and $R = 3$ were empirically determined to produce best results [26]. For each target speaker, UP-AVR thus produced 7 target supervectors ($3 \times 2 + 1$) and 1,065 background supervectors ($355 \times (2+1)$). The new set of labelled supervectors were subsequently used for training speaker-specific SVMs and evaluation, as discussed in Sects. 5.1.1 and 5.1.6, respectively.

Table 5.6 summarizes the effect of UP-AVR on the performances of the GMM-SVM based SV systems in uniform background environments. Drastic performance improvements are noticed compared to the initial set of results (refer Table 5.1). The average EER decrements across all three SNR levels, are 5.13, 6.64, 6.07 and 5.69 % for GMM-SVM (KL div) and 4.79, 6.57, 6.61 and 5.87 % for GMM-SVM (GUMI) in car, factory, pink and white noisy backgrounds, respectively. The magnitude of EER and MinDCF reductions are scaled considerably, thus resolving much of the inconsistencies noted earlier. In contrast to the fractional changes observed initially (see Table 5.1), performance improvements in factory noise backgrounds are observed to be the highest. The average EER improvements across all four types of noises are 7.35, 9.12 and 7.09 % for GMM-SVM (KL div) and 7.04, 9.44 and 7.35 % for GMM-SVM (GUMI) at 0, 5 and 10 dB SNRs, respectively. The GUMI kernels are observed to perform consistently better than the KL div kernels thereby asserting the significance of using covariance information for SV in degraded conditions. However, it is interesting to note that the performance

**Table 5.6** Performance of the GMM-SVM based SV systems with UP-AVR in uniform background environments

| | | GMM-SVM | | | | GMM-SVM with UP-AVR | | | |
|---|---|---|---|---|---|---|---|---|---|
| | | KL div | | GUMI | | KL div | | GUMI | |
| SNR | Noises | EER (%) | MinDCF | EER (%) | MinDCF | EER (%) | MinDCF | EER (%) | MinDCF |
| 0 dB | Car | 14.32 | 0.063 | 13.82 | 0.058 | 11.25 | 0.043 | 11.02 | 0.042 |
| | Factory | 21.27 | 0.086 | 20.32 | 0.086 | 15.36 | 0.063 | 14.54 | 0.059 |
| | Pink | 21.23 | 0.087 | 19.92 | 0.085 | 14.86 | 0.061 | 14.36 | 0.059 |
| | White | 22.72 | 0.076 | 22.67 | 0.079 | 16.03 | 0.066 | 15.67 | 0.063 |
| 5 dB | Car | 13.77 | 0.059 | 13.10 | 0.058 | 07.09 | 0.032 | 06.68 | 0.030 |
| | Factory | 20.87 | 0.085 | 19.92 | 0.084 | 12.24 | 0.048 | 11.74 | 0.047 |
| | Pink | 19.65 | 0.086 | 19.06 | 0.085 | 13.19 | 0.054 | 11.79 | 0.046 |
| | White | 20.96 | 0.074 | 20.73 | 0.072 | 15.36 | 0.065 | 14.27 | 0.061 |
| 10 dB | Car | 12.24 | 0.047 | 11.38 | 0.046 | 06.59 | 0.031 | 06.23 | 0.029 |
| | Factory | 14.81 | 0.053 | 14.50 | 0.052 | 09.44 | 0.039 | 08.76 | 0.038 |
| | Pink | 15.67 | 0.058 | 15.72 | 0.057 | 10.21 | 0.043 | 08.72 | 0.037 |
| | White | 17.75 | 0.065 | 15.49 | 0.067 | 12.96 | 0.055 | 11.33 | 0.048 |
| | Clean | 06.72 | 0.030 | 06.44 | 0.030 | 06.54 | 0.028 | 06.21 | 0.027 |

improvements due to UP-AVR in clean conditions are negligible which explains its effectiveness specifically for noisy backgrounds.

Figure 5.7 demonstrates the impact of UP-AVR in the DET plots of the GMM-SVM based SV systems in uniform noisy environments. A set of red and black lines has been used to denote the upgraded SV systems with KL div and GUMI kernels, respectively. The red and black curves can be easily distinguished from the set of blue and green curves which represents the initial set of GMM-SVM based systems. There is a wide margin of difference at all operating points of the new set of curves in comparison to the old ones. In most cases they are either entirely non-overlapping with the older ones or display the characteristic anti-clockwise rotation. A notable aspect of the UP-AVR based systems is that the performance upgradation at 0 dB SNR is comparable or even better than the initial systems at 10 dB SNR. A comparison of average MinDCF values across all 12 background environments in Table 5.6 show drastic improvements of $19 \times 10^{-3}$ and $22.5 \times 10^{-3}$ for GMM-SVM (KL div) and GMM-SVM (GUMI), respectively.

Table 5.7 summarizes the relative equal error rates of the GMM-SVM based SV systems developed using partitioned utterances in uniform noisy environment. The performance improvements due to UP-AVR are reflected by the dramatic increase in relative EERs. The average relative EERs in car, factory, pink and white noisy backgrounds are 51.94, 39.30, 44.76, 44.36 % for GMM-SVM (KL div) and 53.89, 42.65, 50.01, 48.59 % for GMM-SVM (GUMI), respectively. This is significantly higher than the initial set of relative EERs recorded in Table 5.2. The average improvements in relative EERs are 27.85 and 28.11 % for GMM-SVM (KL div) and GMM-SVM (GUMI), respectively.

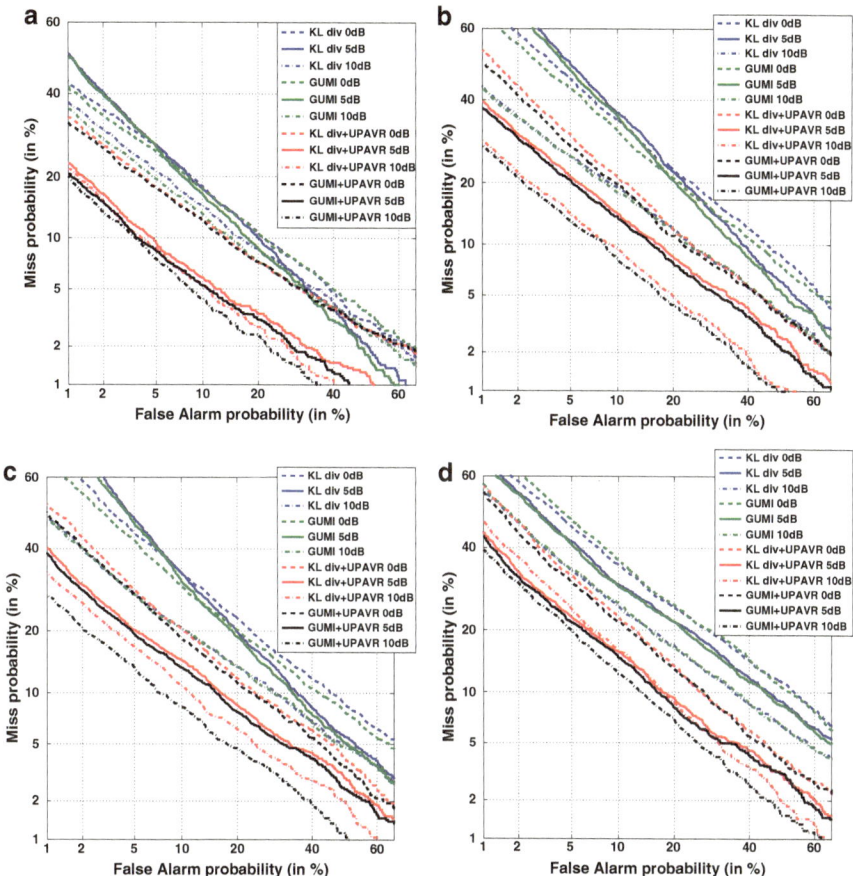

**Fig. 5.7**   DET plots showing the effect of UP-AVR on GMM-SVM based SV systems in uniform background environments with (**a**) car (**b**) factory (**c**) pink and (**d**) white noise

Figure 5.8 demonstrates the changes in (a) EERs and (b) Relative EERs of the GMM-SVM systems with UP-AVR, at various SNRs in uniform background environment. The characteristics of the EERs are in contrast to that observed earlier in Fig. 5.3. Specifically, the abrupt EER fluctuation at 5 dB SNR for factory and pink noises are much relaxed. However close resemblances (with Fig. 5.3) in the relative EER characteristics are noticed with an exception in case of factory noise. As usual an abrupt change in relative EER at 5 dB SNR is noticed with an exception in case of pink and white noise for GMM-SVM (KL div) where linearity in changes are retained.

The effect of UP-AVR was also studied for the SV systems developed in varying background environments. Just like the uniform background scenarios, significant

**Table 5.7** Relative equal error rates for GMM-SVM based SV systems with UPAVR in uniform background environments

| | Relative equal error rate $EER_R$ (%) | | | | | |
|---|---|---|---|---|---|---|
| | SNR (0 dB) | | SNR (5 dB) | | SNR (10 dB) | |
| Noises | KL div | GUMI | KL div | GUMI | KL div | GUMI |
| Car | 37.63 | 38.91 | 60.85 | 63.11 | 57.32 | 59.65 |
| Factory | 33.70 | 37.25 | 41.60 | 43.99 | 42.58 | 46.72 |
| Pink | 44.24 | 46.12 | 44.79 | 50.65 | 45.25 | 53.24 |
| White | 48.26 | 49.42 | 43.96 | 47.94 | 40.85 | 48.29 |

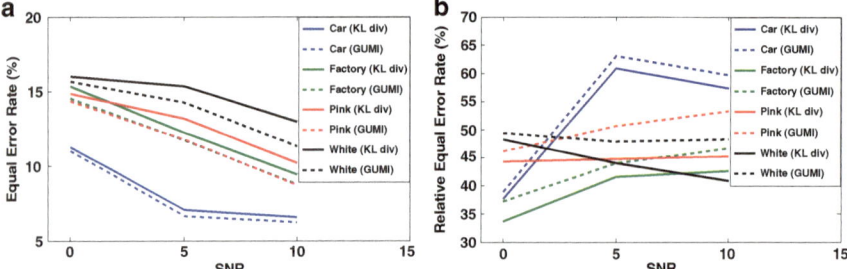

**Fig. 5.8**  (**a**) Equal error rates and (**b**) relative equal error rates of the GMM-SVM based SV systems with UP-AVR at different SNRs in uniform background environments

**Table 5.8** Performance of the GMM-SVM based SV systems with UP-AVR in varying background environments

| | GMM-SVM | | | | GMM-SVM with UP-AVR | | | |
|---|---|---|---|---|---|---|---|---|
| SNR | KL div | | GUMI | | KL div | | GUMI | |
| (dB) | EER (%) | MinDCF | EER (%) | MinDCF | EER (%) | MinDCF | EER (%) | MinDCF |
| 0 | 23.48 | 0.086 | 22.76 | 0.085 | 15.76 | 0.060 | 16.16 | 0.066 |
| 5 | 22.18 | 0.080 | 21.32 | 0.084 | 15.18 | 0.053 | 14.81 | 0.052 |
| 7 | 19.74 | 0.073 | 19.11 | 0.071 | 14.50 | 0.051 | 12.38 | 0.048 |
| 10 | 18.65 | 0.072 | 16.44 | 0.069 | 12.24 | 0.047 | 11.38 | 0.046 |

reduction in the error metrics are observed once again, in contrast to the initial set of system performances (without UP-AVR), as shown in Table 5.8. The EER reductions compared to the initial set of observations are 7.72, 7.00, 5.24, 6.41 % for GMM-SVM (KL div) and 6.60, 6.51, 6.73, 5.06 % for GMM-SVM (GUMI) at 0, 5, 7 and 10 dB SNRs, respectively. The two SVM kernels show different behavior in terms of EER changes with a slight anomaly noticed at 0 dB SNR where the KL div kernel performs better than the GUMI kernel. The effect of UP-AVR also appears to be more prominent in case of KL div kernels which shows an average EER improvement of 6.59 % across all SNR levels in comparison to 6.23 % for the GUMI kernel.

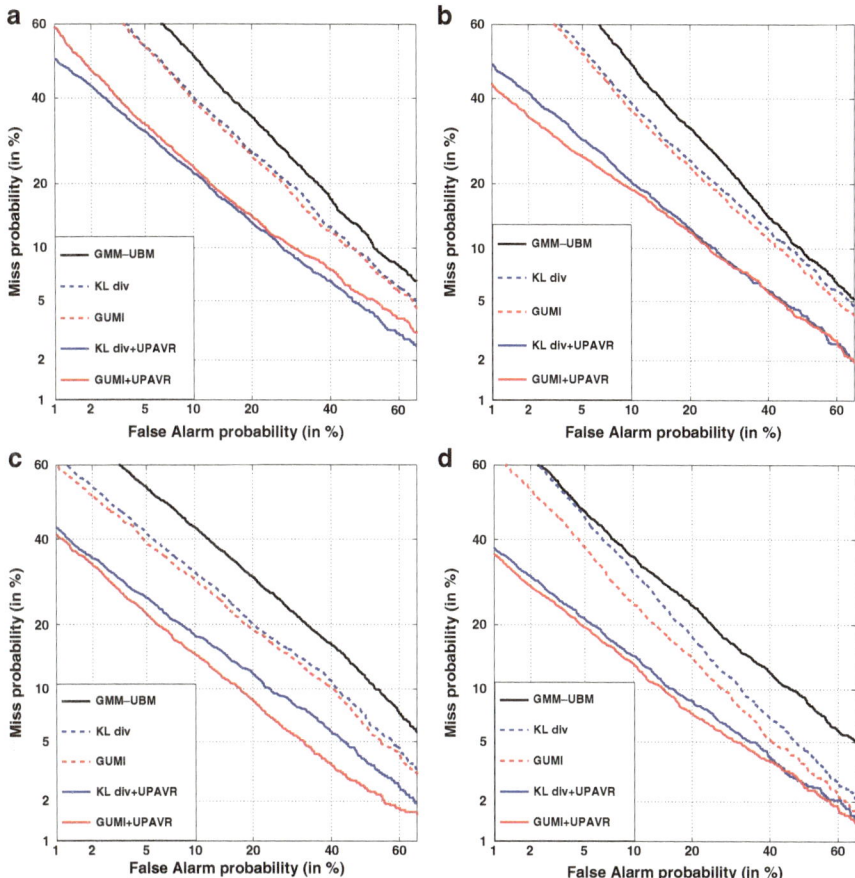

**Fig. 5.9** DET plots showing the effect of UP-AVR on GMM-SVM based SV systems in varying background environments at (**a**) 0 dB (**b**) 5 dB (**c**) 7 dB and (**d**) 10 dB SNRs

Figure 5.9 demonstrates the impact of UP-AVR in the DET plots of the GMM-SVM based SV systems in varying background environments. The set of solid blue and red lines denote the UP-AVR based GMM-SVM systems with KL div and GUMI kernels, respectively. The broken lines of same colors represent the initial systems developed in the same backgrounds while the black line represents the baseline. As usual a wide margin is noticed between the solid and broken set of curves. Unlike the initial set of GMM-SVM systems (see Fig. 5.4), dissimilarities are observed in the curves corresponding to the two SVM kernels. In most cases, the red and blue curves show distinct behavior. Apart from the overall shift towards the origin, anti-clockwise rotations in the red and blue curves are prominently noticed in

**Table 5.9** Performance of the GMM-SVM based SV systems with UP-AVR and NAP compensation in varying background environments

| SNR (dB) | GMM-SVM + UP-AVR | | | | GMM-SVM + UP-AVR + NAP | | | |
|---|---|---|---|---|---|---|---|---|
| | KL div | | GUMI | | KL div | | GUMI | |
| | EER (%) | MinDCF | EER (%) | MinDCF | EER (%) | MinDCF | EER (%) | MinDCF |
| 0 | 15.76 | 0.060 | 16.16 | 0.066 | 13.37 | 0.058 | 13.62 | 0.063 |
| 5 | 15.18 | 0.053 | 14.81 | 0.052 | 13.23 | 0.051 | 12.47 | 0.048 |
| 7 | 14.50 | 0.051 | 12.38 | 0.049 | 13.10 | 0.049 | 11.29 | 0.046 |
| 10 | 12.24 | 0.047 | 11.38 | 0.046 | 11.21 | 0.044 | 10.32 | 0.043 |

case of 5 and 10 dB SNRs. The resultant improvement in MinDCF values averaged across all SNRs are $25.00 \times 10^{-3}$ and $24.25 \times 10^{-3}$ for KL div and GUMI kernels, respectively.

As mentioned earlier in Sect. 5.2, the UP-AVR strategy was adopted to alleviate a set of three highlighted drawbacks of the conventional GMM-SVM based systems. However, it is difficult to conclude the degree of impact UP-AVR exercises on each of them. In most cases one may rely on the joint improvement of all three problems, without specifically knowing each of them. In order to demonstrate the specific utility of UP-AVR towards mitigating the small sample-size problem, the partitioned enrollment utterances (supervectors) were subjected to NAP transformation prior to SVM training.

Unlike its earlier version, the supervector matrix $V$ constructed in the Step 3 of the NAP algorithm, now had an expanded size of $2,800 \times 39,936$ due to the impact of UP-AVR on the target speaker utterances. All the training supervectors were subjected to NAP transformation prior to SVM training with required strategies for maintaining feasibility in large matrix operations as discussed in Sect. 5.1.6.1.

Table 5.9 summarizes the performance of the GMM-SVM based SV developed using UP-AVR followed by NAP compensation in varying background environments. In contrast to the initial set of observations (see Table 5.4), a larger average EER reduction compared to the baseline (i.e., 12.26 % (KL div) and 13.06 % (GUMI)) is noticed across all four SNR levels. The additional improvements due to NAP over UP-AVR are 2.39, 1.95, 1.40, 1.03 % and 2.54, 2.34, 1.09, 1.06 % at 0, 5, 7 and 10 dB SNRs for GMM-SVM (KL div) and GMM-SVM (GUMI), respectively. The effect of NAP compensation is observed to be more prominent in case of the GUMI kernels.

Figure 5.10 demonstrates the effect of NAP on the DET plots of the GMM-SVM based systems developed in varying background environments. The color coding for representing each system is the same as that in Fig. 5.9. The broken lines of each color have been used to denote the corresponding systems with NAP transformation. The set of blue and green lines seen earlier in Fig. 5.5, are included again for studying the overall comparison of the various systems. The behaviour of the NAP-based curves are quite similar in shape and alignment to the initial systems except for a larger margin of difference from them at all operating points. Expectedly, a significant improvement in MinDCF values are observed

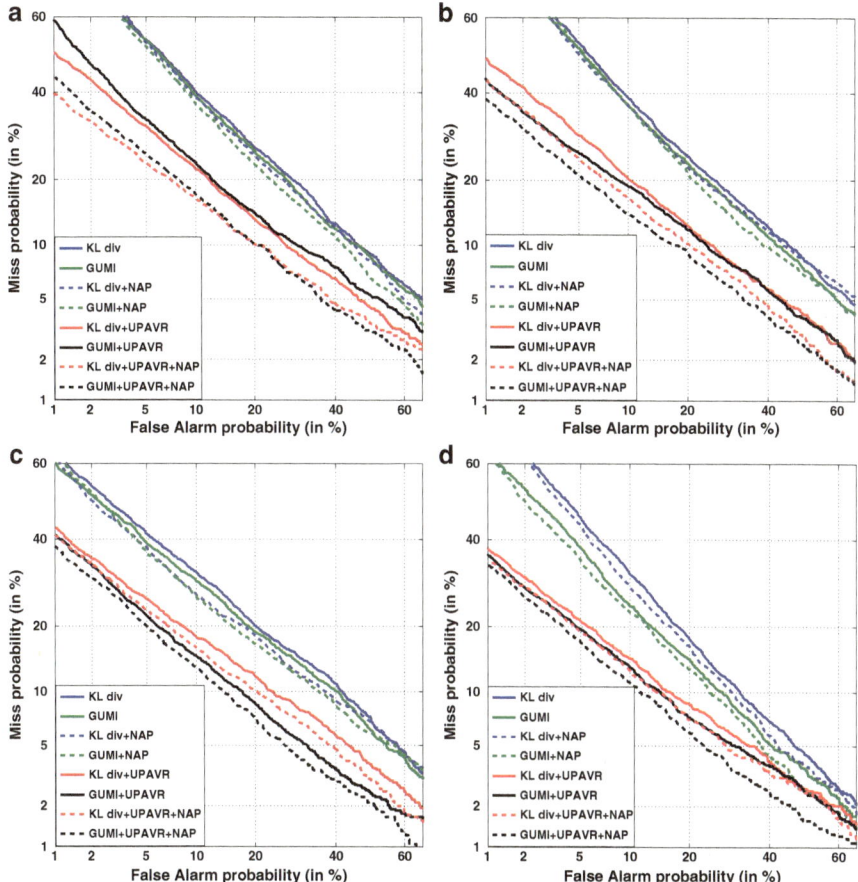

**Fig. 5.10**  DET plots showing the effect of UP-AVR and NAP on GMM-SVM based SV systems in varying background environment at (**a**) 0 dB (**b**) 5 dB (**c**) 7 dB and (**d**) 10 dB SNRs

in comparison to the earlier NAP-based SV systems with average reduction of $26.00 \times 10^{-3}$ and $25.75 \times 10^{-3}$ across all SNRs, for KL div and GUMI kernels, respectively.

Table 5.10 summarizes the relative EERs of the various GMM-SVM based SV systems developed in varying background environments. The average relative EERs across all SNRs, are 42.36 and 45.43 % for UP-AVR based GMM-SVM systems with KL div and GUMI kernels, respectively. The corresponding values with an additional NAP application are 49.02 and 52.34 %. The benefits of utterance partitioning can be observed from the significant average improvements of 26.68 and 25.45 % relative EER rates for the two types of NAP based GMM-SVM based systems.

**Table 5.10**  Comparison of relative equal error rates for GMM-SVM based SV systems in varying background environments

| SNR (dB) | Relative equal error rate $EER_R$ (%) | | | | | | | |
| --- | --- | --- | --- | --- | --- | --- | --- | --- |
| | GMM-SVM | | NAP | | UP-AVR | | UP-AVR + NAP | |
| | KL div | GUMI | KL div | GUMI | KL div | GUMI | KL div | GUMI |
| 0 | 13.20 | 15.86 | 17.86 | 21.37 | 41.74 | 40.26 | 50.57 | 49.65 |
| 5 | 13.83 | 17.17 | 18.22 | 21.76 | 41.03 | 42.46 | 48.60 | 51.55 |
| 7 | 21.95 | 24.44 | 25.90 | 28.59 | 42.67 | 51.05 | 48.20 | 55.36 |
| 10 | 14.68 | 24.79 | 18.80 | 27.86 | 44.01 | 47.94 | 48.72 | 52.79 |

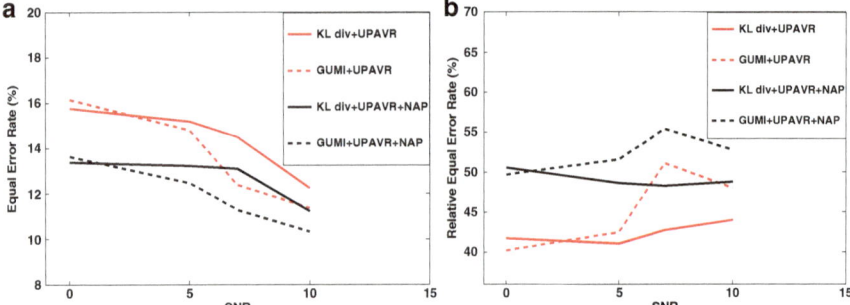

**Fig. 5.11**   (**a**) Equal error rates and (**b**) relative equal error rates of the GMM-SVM based SV systems with UP-AVR at different SNRs in varying background environments

Figure 5.11 demonstrates the changes in (a) EERs and (b) Relative EERs of the GMM-SVM systems with UPAVR, at various SNRs in varying background environments. Notable changes in the behavior of the red and black curves are observed in contrast to those in Fig. 5.6. The abrupt EER and relative EER fluctuations at 7 dB SNR, initially observed in Fig. 5.6 are now relaxed for the GMM-SVM (KL div) system where NAP application makes an anomalous change. The GUMI based GMM-SVM systems however show similar behavior with and without NAP applications which is characterized by consistent increase in relative EER values with increasing SNRs with an abrupt decrement at the 10 dB SNR level.

## 5.3   Total Variability Modeling for Speaker Verification

The significance of the GMM-SVM methods for SV in noisy environment was explored through an extensive set of empirical studies discussed in Sects. 5.1 and 5.2, respectively. Despite the drastic performance enhancements achieved using the UP-AVR strategy, few typical limitations of the developed SV systems can be highlighted. Firstly, the large size of the GMM supervectors are a practical constraint in terms of their memory consumption and computational costs (e.g.,

SVM training, NAP transformation). Secondly, despite UP-AVR the performance improvements of the SV systems developed in extremely degraded conditions in the uniform background environments were comparatively lower. Specifically, the average relative EERs of the SV systems across the four different backgrounds were 41.95 and 49.61 % at 0 and 5 dB SNRs in contrast to a larger value of 54.23 % at 10 dB SNR. Even the average EERs at 0 and 5 dB i.e., 14.13 and 11.54 % were significantly larger than those at 10 dB (9.28 %). Individual EERs were observed to be typically high for factory, pink and white noisy backgrounds. These factors suggested the use of alternative robust speaker modeling methods for further improvement in performance accuracy. Specifically a state-of-the-art low dimensional representation of GMM supervectors, commonly known as identity vectors or 'i-vectors' [7], was used for developing SV systems. In the remaining part of this section, the details of i-vector extraction, its application and evaluation in the present work are discussed.

### 5.3.1  i-Vector Extraction

Total variability modeling [7] is based on projecting large dimensional supervectors in a low dimensional subspace (known as 'total variability' space) which supposedly contains both channel and session information. Specifically, a GMM mean super-vector $M$ is represented as

$$M = m + Tw \tag{5.11}$$

where $m$ is a speaker/channel independent supervector (i.e., the UBM mean super-vector), $T$ is low-rank rectangular matrix whose columns consists of eigenvectors of the total variability covariance matrix with largest eigenvalues. $w$ is a random vector having standard Normal distribution, called i-vector. The total variability matrix $(T)$ is learned offline, using probabilistic principal component analysis (PPCA) [4] on a development dataset [27, 28]. Estimation of i-vectors from a set of utterances requires initial computation of a set of Baum-Welch statistics followed by a set of matrix operations involving them. Given a sequence of $D$-dimensional acoustic vectors $\{x_1, x_2, \ldots x_T\}$ of an utterance $X$ with $\mathbf{T}$ frames, the Baum-Welch statistics are calculated as

$$N_i = \sum_{t=1}^{\mathbf{T}} p(i|x_t, \lambda)$$

$$F_i = \sum_{t=1}^{\mathbf{T}} p(i|x_t, \lambda)(x_t - m_i)$$

where $p(i|x_t, \lambda)$ is the posterior probability of the $i$th Gaussian component of a UBM $\lambda$ having total $M$ components, which generates vector $x_t$. The mean of the same component is given by $m_i$. $N_i$ and $F_i$ are known as the zeroth order and mean-shifted first order sufficient statistics, respectively.

Given the trained $T$ matrix and the set of Baum-Welch statistics, the i-vector extracted from utterance $X$ is calculated as

$$w = (I + T^T \Sigma^{-1} N(X) T)^{-1} . T^T \Sigma^{-1} F(X) \qquad (5.12)$$

where $\Sigma$ and $N(X)$ are block diagonal matrices of size $(MD \times MD)$ whose diagonal blocks consist of the UBM covariance matrices $\Sigma_i$ $(i = 1, 2, \ldots M)$ and identity matrices weighted with the zeroth order statistics $N_i I_{D \times D} (i = 1, 2, \ldots M)$, respectively. F(X) is a supervector obtained by stacking the mean-shifted first order statistics $F_i$ $(i = 1, 2, \ldots M)$ and $I$ is an identity matrix of size $(MD \times MD)$. Total variability modeling is generative in nature, however they can be integrated with a discriminative framework using SVMs [29, 30]. The detailed procedure of training the $T$ matrix has been outlined in Appendix D.

### 5.3.2   SVM Training

Since i-vectors are fixed-length vectors representing variable length utterances, they can be used to train SVMs using sequence kernels as discussed in Sect. 5.1.1. It was investigated in [29], that the best result in i-vector frameworks are produced by using a cosine kernel function for training the SVMs, which can be defined for two input i-vectors $w_1$ and $w_2$ as

$$k(w_1, w_2) = \frac{\langle w_1, w_2 \rangle}{\|w_1\| \|w_2\|} \qquad (5.13)$$

where $\langle ., . \rangle$ and $\|.\|$ denote the inner product and L2-norm, respectively. The cosine kernel normalizes the linear kernel by the norm of both i-vectors. It considers only the angle between the two i-vectors and not their magnitudes. It is believed that non-speaker information (such as session and channel) affects the i-vector magnitudes, removing which improves the robustness of the i-vector system.

### 5.3.3   Inter-session Compensation

Since the total variability subspace contains both speaker and session variability information, i-vectors extracted from it are usually subjected to session compensation prior to SVM training. Two common session compensation techniques used in the i-vector framework are discussed as follows

### 5.3.3.1 Linear Discriminant Analysis (LDA)

LDA [4] projects the i-vectors to a set of orthogonal axes for minimizing within-class variance and maximizing between-class variance. In the i-vector framework, all i-vectors extracted from a speaker constitute a particular class. The projection matrix $A$ is composed of eigenvectors $v$ having the highest eigenvalues $\lambda$, obtained by solving the following generalized eigen decomposition problem

$$B_S v = \lambda W_S v \tag{5.14}$$

where $B_S$, $W_S$ are the between-class and within-class covariance matrices given by

$$B_S = \sum_{s=1}^{S} (\mu_s - \mu)(\mu_s - \mu)^T \tag{5.15}$$

$$W_S = \sum_{s=1}^{S} \frac{1}{n_s} \sum_{i=1}^{n_s} (w_i - \mu_s)(w_i - \mu_s)^T \tag{5.16}$$

where $S$ is the total number of speakers, $n_s$ is the total number of utterances from the $s$th speaker, $\mu_s$ is the mean of all i-vectors $(w_i)$ from speaker $s$ given by $\mu_s = \frac{1}{n_s} \sum_{i=1}^{n_s} w_i$ and $\mu$ is the global mean of all i-vectors generally considered to be a null vector due to their standard normal distribution. The number of columns of the matrix $A$ (i.e., LDA order) are determined empirically to produce best results. The LDA-modified cosine kernel function for two input i-vectors $w_1$ and $w_2$ is given by

$$k(w_1, w_2) = \frac{(A^T w_1)^T (A^T w_2)}{|A^T w_1||A^T w_2|} \tag{5.17}$$

### 5.3.3.2 Within-Class Covariance Normalization (WCCN)

WCCN, proposed in [31] aims to set upper bounds on the error metrics ('miss' and 'false alarm') by normalizing the SVM kernels. Application of WCCN in the i-vector framework requires projecting the i-vectors to a space specified by the square-root of the inverse of the within-class covariance matrix. Specifically, the projection matrix $B$ is obtained by Cholesky decomposition of the inverse of the within-class covariance matrix (Eq. 5.16) as follows

$$W_S^{-1} = BB^T \tag{5.18}$$

The WCCN-modified cosine kernel function for two input i-vectors $w_1$ and $w_2$ is given by

$$k(w_1, w_2) = \frac{(B^T w_1)^T (B^T w_2)}{|B^T w_1||B^T w_2|} \tag{5.19}$$

### 5.3.4 Score Calculation

Two types of i-vector evaluation methods, namely Cosine Distance scoring and
SVM Kernel scoring, has been proposed in past [29]. The former one is applied
in the default generative modeling framework while the latter for the discriminative
SVM framework.

#### 5.3.4.1 Cosine Distance Scoring (CDS)

CDS is a fast scoring method commonly applied in i-vector frameworks. As the
name suggests, it is simply the cosine distance between a pair of i-vectors representing a claimant's test utterance ($w^{test}$) and the claimed target speaker utterance
($w^{target}$), respectively as given by

$$S_{cos} = \frac{< w^{test}, w^{target} >}{|w^{test}||w^{target}|} \qquad (5.20)$$

#### 5.3.4.2 SVM Kernel Scoring

The SVM scoring is exactly similar to the one already discussed in Sect. 5.1.4. The
advantage of SVM scoring is that the contribution of individual speakers towards
the verification scores can be optimally weighted by the Lagrange multipliers of the
target speakers SVM. Given a trained target speaker SVM and the test i-vector $w^{test}$,
the score is calculated as

$$S_{SVM} = \sum_{t=1}^{T} \alpha_t K(w^t, w^{test}) - \sum_{i=1}^{B} \alpha_i K(w^i, w^{test}) + d \qquad (5.21)$$

where $w^t$ and $w^i$ are the sequence of support vectors corresponding to the target
and background speaker classes as learned during SVM training. $\alpha_t$ and $\alpha_i$ are the
non-zero Lagrange multipliers of the corresponding classes. $T$ and $B$ are the total
number of support vectors in each class, $d$ is a bias term and $K$ is the cosine kernel
(Eq. 5.13).

### 5.3.5 Experimental Setup

Figure 5.12 shows a block diagram of the i-vector based SV system. The SV systems
were developed using the set of noise-degraded training and test utterances of
NIST-SRE-2003 in uniform background environment (see Sect. 5.1.5.1) at 0 and
5 dB SNRs, as discussed in Sects. 5.3.1 and 5.3.3. A development data comprising

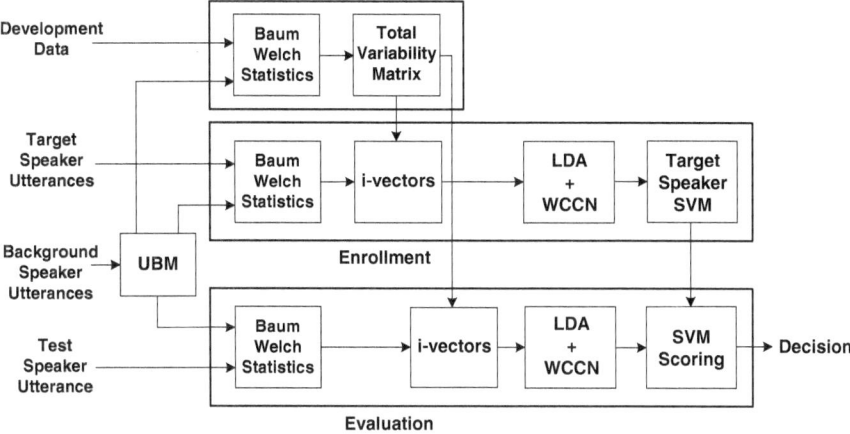

**Fig. 5.12** Block diagram of the combined SVM and total variability modeling framework for speaker verification

1,572 utterances from the SwitchBoard phase II corpora and 400 utterances from NIST-SRE-2004 database was used for training the total variability and channel compensation matrices (see Appendix D). The $T$-matrix rank of 400 was chosen empirically and i-vectors were extracted from all utterances as discussed in Sect. 5.3.1. The low dimension of i-vectors facilitated convenient application of LDA and WCCN, the projection matrices for which were designed as discussed in Sect. 5.3.3. A LDA order of 300, was empirically determined to produce best results. All the i-vectors were subjected to session compensation prior to model building. A discriminative framework (combined i-vector and SVM) for classification was used instead of the conventional generative i-vector modeling in favor of utilizing the benefits of UP-AVR and SVM scoring as shown in [15]. The labelled i-vectors extracted from enrollment and background speaker's utterances were subjected to speaker specific SVM training. During the evaluation phase, the noisy test utterances (i-vectors) were evaluated against the target speaker models (SVMs) according to NIST-2003 primary task, using the SVM scoring method as discussed in Sects. 5.1.6 and 5.3.4.2, respectively. The experiments were repeated using partitioned utterances with UP-AVR parameters $N = 2$ and $R = 3$.

Table 5.11 summarizes the performance of the i-vector based SV systems developed in uniform noisy environments at 0 and 5 dB SNR. While the error metrics show considerable performance improvements compared to the GMM-SVM based systems in individual noisy backgrounds, it is interesting to note that the GMM-SVM based systems developed using UP-AVR performs better than the i-vector models developed without UP-AVR. This can be deduced from a comparison of the GMM-SVM based SV systems in (Table 5.6). The average EER reductions across both SNRs compared to the default GMM-SVM based SV systems are 3.52, 6.12, 4.90 and 5.51% for car, factory, pink and white noisy backgrounds,

**Table 5.11** Performance of the i-vector based SV systems in uniform background environments at 0 and 5 dB SNR

| Noises | SNR (0 dB) | | | | SNR (5 dB) | | | |
|---|---|---|---|---|---|---|---|---|
| | Without UP-AVR | | With UP-AVR | | Without UP-AVR | | With UP-AVR | |
| | EER (%) | MinDCF | EER (%) | MinDCF | EER (%) | MinDCF | EER (%) | MinDCF |
| Car | 12.10 | 0.051 | 10.57 | 0.047 | 08.31 | 0.039 | 06.05 | 0.028 |
| Factory | 16.17 | 0.068 | 12.33 | 0.055 | 12.78 | 0.057 | 09.03 | 0.042 |
| Pink | 16.72 | 0.071 | 12.83 | 0.054 | 13.42 | 0.058 | 10.16 | 0.044 |
| White | 17.39 | 0.073 | 14.27 | 0.059 | 15.13 | 0.063 | 11.88 | 0.051 |

respectively. However, GMM-SVM with UP-AVR performs slightly better than the current systems with average reduced EERs of 1.23, 1.01, 1.52 and 0.93% at the corresponding environments across both SNRs. This phenomenon once again establishes the significance of UP-AVR in enhancing SV performances in noisy conditions. The superiority in i-vector performance accuracies are restored by incorporating UP-AVR in its framework. Comparison amongst the UP-AVR based systems (see Table 5.6) reveals average EER reductions of 0.70, 2.79, 2.06, 2.26% in car, factory, pink and white noisy backgrounds, respectively.

Figure 5.13 shows the DET plots of the i-vector based SV systems developed in uniform noisy environments. As usual a shift towards the origin is observed in the curves corresponding to the UP-AVR based systems (represented by broken lines) suggesting consistent reduction in MinDCF and EER values across each noisy background. The effect of UP-AVR at 0 dB SNR is apparently more prominent in case of the colored noises. Unlike the GMM-SVM based systems (see Fig. 5.7), no significant change in slope or rotation of the curves are noticed. The average improvements in MinDCF values of the i-vector based SV systems (with and without UP-AVR) in comparison to the corresponding GMM-SVM based SV systems (see Tables 5.1 and 5.6) are $2.94 \times 10^{-3}$ and $1.82 \times 10^{-3}$, respectively.

Despite the apparent performance improvements achieved by the i-vector based SV systems, a typical aspect to be noticed is that UP-AVR results in a moderate decrement of only 3.10 % average EER. Similar observations were earlier recorded for the GMM-SVM based systems (see Table 5.6) which had shown 5.39 % EER reductions at 0 dB SNR in contrast to larger improvements at 5 and 10 dB SNRs. This phenomenon indicates the obvious increase in classification errors due to high noise strength. A typical drawback of the standard UP-AVR algorithm can also be highlighted in this context. Specifically, all speaker's utterances are partitioned irrespective of the role they play towards classification. This could be detrimental towards SV performance e.g., partitioning a speaker's utterance which was originally misclassified could lead to additional misclassifications apart from increased computational load. In order to alleviate these two problems in parallel, a novel boosting algorithm is proposed to train multiple SVM classifiers on the noisy dataset, the utterances in which are selectively used for partitioning. Subsequent sections provide the details of the boosting algorithm followed by their implementation in the i-vector based SV framework.

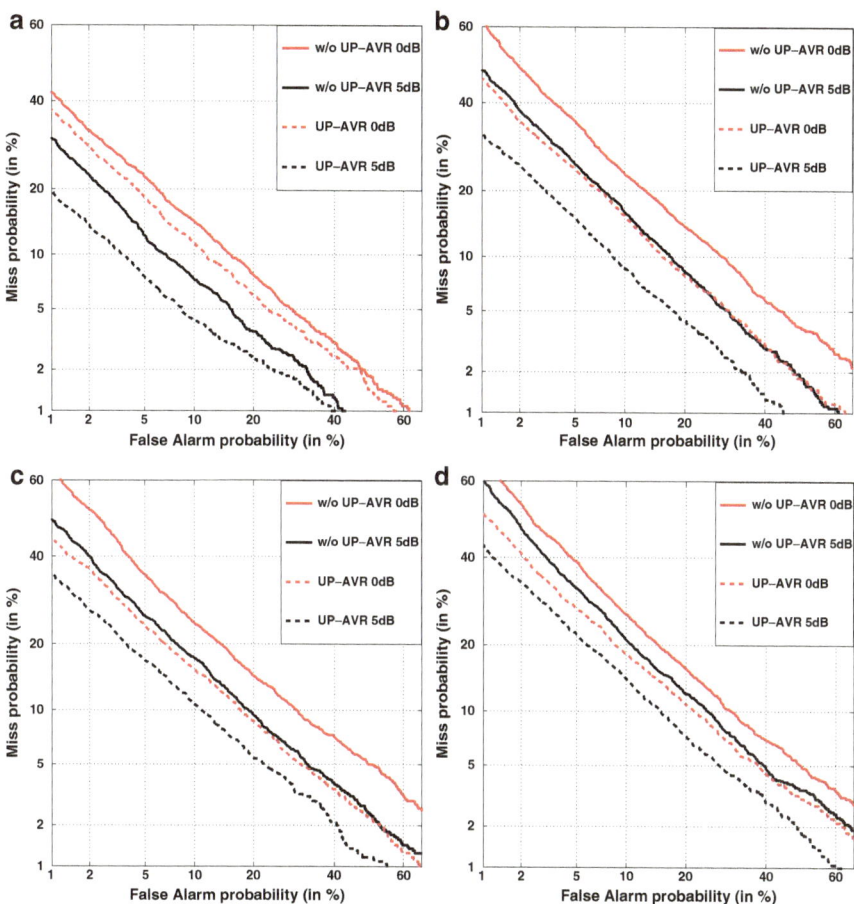

**Fig. 5.13** DET plots of the *i-vector* based SV systems in uniform background environments with (**a**) car (**b**) factory (**c**) pink and (**d**) white noise at 0 and 5 dB SNR

## 5.4 Adaptive Boosting for Improved Speaker Verification Performance in Noisy Environments

Adaptive Boosting (AdaBoost) iteratively enhances the predictive accuracy of a sequence of weak classifiers (ensemble), each of which is trained on a dataset adaptively sampled according to the training error of the classifier in the previous iteration [32]. The final decision is based on a weighted voting of the individual classifiers in the ensemble. In recent past, boosting has been applied effectively for robust SV tasks [33]. Prior art also demonstrates the benefits of combining ensemble learning with data balancing [20, 34]. A novel combination scheme of the AdaBoost algorithm with a synthetic data generation technique using UP-AVR

[25], is proposed in the present work. The approach is motivated by the Databoost-IM algorithm proposed in [35]. The aim is to improve the predictive accuracy of both minority (target speaker) and majority (background speakers) classes while emphasizing on the misclassified examples in the minority class.

### 5.4.1  Proposed Boosting Algorithm (DataBoost-UP)

Conventional boosting algorithms emphasize on the misclassified (hard) training instances at each iteration by adaptively increasing their sampling weights. Classifiers trained in successive iterations concentrate on these instances with high weights. Since all misclassified examples are equally weighted, it doesn't compensate for the bias towards the majority class in imbalanced datasets. The aim of integrating data generation with the boosting algorithm is to alleviate the learning algorithm's bias towards the majority class while retaining focus on the hard training instances. Unlike the DataBoost-IM algorithm [35], in the proposed algorithm (DataBoost-UP) the data (i-vectors) is synthesized using the utterance partitioning technique [25] instead of random generation of attribute values in the [min,max] interval. Both the minority (target speaker) and majority (background speakers) classes are oversampled to prevent overemphasis on the hard instances of the minority class. The proposed algorithm is used to create an ensemble of SVM classifiers.

---

**Algorithm**  DataBoost-UP

---

**Input:**

   Training data set $\{(x_i, y_i)\}_{i=1}^{N}$, $y_i \in \{-1, +1\}$

   Weak SVM classifiers $h_t$ where $t = \{1, 2 \ldots, T\}$

**Initialize:** Sampling weight distribution $D_1(i) = 1/N \ \ \forall i = \{1, 2, .., N\}$

**Do for t** $\leftarrow$ 1 to T

1. Identify the hard examples in the training set.
2. Generate new data from these examples by UP-AVR. Add them to the original training set.
3. Adjust the sampling weight distribution of both classes in the new training set.
4. Learn weak SVM $h_t$ on the new training set sampled according to the modified distribution.
5. $\epsilon_t \leftarrow \sum\limits_{i=1}^{N} D_t(i) I(h_t(x_i) \neq y_i)$. If $\epsilon_t > 0.5$ set T = t-1 and abort loop.
6. $\alpha_t \leftarrow \frac{1}{2} \log\{(1 - \epsilon_t)/(\epsilon_t)\}$
7. $D_{t+1}(i) \leftarrow \frac{D_t(i)}{Z_t} \exp(-\alpha_t h_t(x_i) y_i)$ where $Z_t = \sum\limits_{i=1}^{N} D_t(i) \exp(-\alpha_t h_t(x_i) y_i)$

**Output:** SVM ensemble $h_{final} = \sum\limits_{t=1}^{T} \alpha_t h_t$

---

The predictive accuracy of the ensemble is guaranteed to improve in each iteration provided the training error of the weak SVM classifier in the previous iteration is less than 0.5 (upper bound). The ensemble training error decreases in successive iterations. At the end of a pre-determined number of iterations, the algorithm converges with no further decrement in the ensemble training error. Steps 1, 2 and 3 of the proposed algorithm are elaborated in the next three sections.

### 5.4.1.1  Identifying Hard Training Examples

The hard training examples are identified as follows.

1. All the instances in the training set are arranged in descending order of their sampling weights.
2. The top $N_{train}$ number of instances of the training set are selected as hard examples where:
   $N_{train} = \epsilon_t \times N$,
   $\epsilon_t =$ weighted training error of a SVM in the $t$th iteration of boosting
   $N =$ total number of instances in the original training set.
3. Let $N_{train} = N_{maj} + N_{min}$ where:
   $N_{maj} =$ number of instances from majority class, $N_{min} =$ number of instances from minority class.
   These training utterances are subjected to utterance partitioning as discussed in Sect. 5.4.1.2

### 5.4.1.2  Synthesizing Data Using Utterance Partitioning

The UP-AVR algorithm (discussed in Sect. 5.2) is applied for data generation, as follows

1. Given each of the $N_{min}$ target speaker utterance, its acoustic vectors are computed and their sequence of occurrences in the utterance are randomized. This randomized sequence is then divided into $P$ partitions (sub-utterances).
2. Step 1 is repeated $R$ times. Together with the original full-length utterance, a total of $RP + 1$ utterances generated from each enrollment utterance are individually subjected to i-vector construction.
3. Similarly, each background speaker's utterances are divided into $P$ partitions. For $N_{maj}$ background speakers we thus have $N_{maj}(P + 1)$ utterances. Background i-vectors are constructed from each of these utterances.

### 5.4.1.3  Balancing Weights of Majority and Minority Classes

The aim of weight balancing is to minimize the difference between the total sampling weight of each class in an imbalanced dataset. This forces the boosting algorithm to focus on both the hard as well as rare training examples. The sampling

**Table 5.12** Comparison of the effects of UP-AVR and Databoost-UP on the performances of i-vector based SV systems in uniform background environments at 0 and 5 dB SNRs

| Noises | SNR (0 dB) | | | | SNR (5 dB) | | | |
| | UP-AVR | | DataBoost-UP | | UP-AVR | | DataBoost-UP | |
| | EER (%) | MinDCF | EER (%) | MinDCF | EER (%) | MinDCF | EER (%) | MinDCF |
|---|---|---|---|---|---|---|---|---|
| Car | 10.57 | 0.047 | 08.22 | 0.037 | 06.05 | 0.028 | 04.83 | 0.021 |
| Factory | 12.33 | 0.055 | 10.93 | 0.048 | 09.03 | 0.042 | 07.14 | 0.032 |
| Pink | 12.83 | 0.054 | 11.21 | 0.047 | 10.16 | 0.044 | 08.13 | 0.035 |
| White | 14.27 | 0.059 | 13.05 | 0.053 | 11.88 | 0.051 | 10.03 | 0.043 |

weight of each hard instance is divided by the number of new instances generated from it. All generated instances are uniformly assigned the divided weight. At the end the weights are rebalanced across the entire set of newly generated instances. If the total weight of the majority class ($W_{maj}$) exceeds that of the minority class ($W_{min}$) then each minority weight is scaled by a factor $W_{maj}/W_{min}$. For the vice-versa condition, each majority weight is scaled by a factor $W_{min}/W_{maj}$.

### 5.4.2  Performance Evaluation

The data used for experimental setup is identical to that described in Sect. 5.3.5. The i-vectors extracted from the partitioned target speaker utterances from each noisy dataset were used for training a SVM ensemble using the DataBoost-UP algorithm. Additionally, new data was generated in each iteration of the boosting algorithm with partitioning parameters values of $P = 2$ and $R = 1$ as discussed in Sect. 5.4.1.2. The number of boosting iterations ranging from 5 to 10 was empirically determined to appropriately lower the ensemble training error. During the evaluation phase, each test utterances were scored against 11 target speaker SVM ensemble. Given a noisy test utterance (i-vector) $w^{test}$, the Kernel scoring was obtained as a weighted linear combination of the scores obtained from individual classifiers of the target speaker ensemble as follows:

$$Score(w^{test}) = \sum_{i=1}^{T} \alpha_i \left( \sum_{j=1}^{L} \beta_{i,j} t_{i,j} K(w^{i,j}, w^{test}) + d_i \right)$$

where $T$ is the size of the ensemble. $\alpha_i$ is the weight of the $i$th SVM classifier in the ensemble as calculated in Step 6 of the DataBoost-UP algorithm. $w^{i,j}$, $\beta_{i,j}$ and $t_{i,j} \in \{-1, +1\}$ are the sequence of $L$ learned support vectors, the non-zero Lagrange multipliers and the actual class labels, respectively for the $i$th SVM classifier in the ensemble, $d_i$ is the bias term and $K$ is the cosine kernel function.

Table 5.12 summarizes the comparative performances of DataBoost-UP and UP-AVR method in the i-vector framework for the SV systems developed in uniform

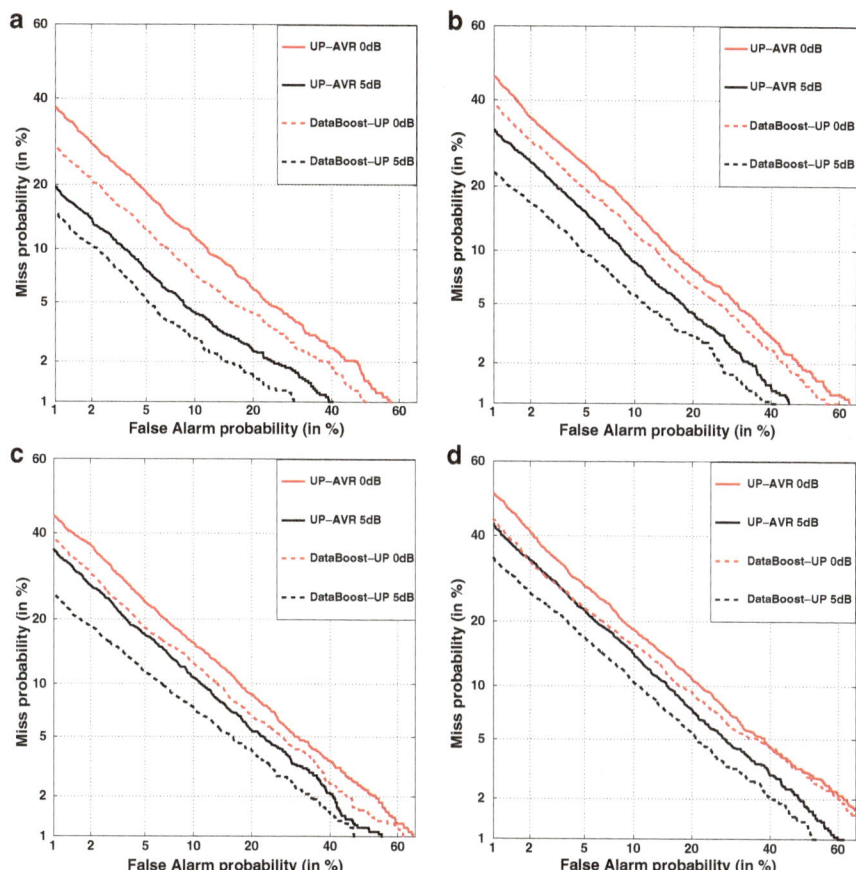

**Fig. 5.14** DET plots showing the effect of UP-AVR and DataBoost-UP on the *i-vector* based SV systems in uniform background environment with (**a**) car (**b**) factory (**c**) pink and (**d**) white noise at 0 and 5 dB SNR

background environments. A consistent performance improvement is noticed in the boosted i-vector framework across all noisy backgrounds, in comparison to the UP-AVR based system at both SNR levels. The individual EER reductions are 2.35, 1.40, 1.62 and 1.22 % at 0 dB SNR and 1.22, 1.89, 2.03, 1.85 % at 5 dB SNR for car, factory, pink and white noisy backgrounds, respectively. Thus an additional average EER reduction of 1.70 % across all environments is achieved on top of the initial improvement (see Table 5.11) of 3.12 % due to UP-AVR.

Figure 5.14 shows the DET plots of the i-vectors based SV systems using (a) UP-AVR and (b) DataBoost-UP, respectively. Interestingly, the nature of improvement in each curve is similar to those observed in the Fig. 5.13. There are no apparent rotation in the curves apart from the overall shift towards the origin characterized by the corresponding changes in detection costs. In contrast

**Table 5.13** Comparison of relative equal error rates of SV systems developed in uniform background environments at 0 and 5 dB SNRs

| SNR (dB) | Noises | Relative equal error rate $EER_R$ (%) | | | | |
|---|---|---|---|---|---|---|
| | | GMM-SVM (supervectors) | | Total variability (i-vectors) | | |
| | | w/o UP-AVR | With UP-AVR | w/o UP-AVR | With UP-AVR | DataBoost-UP |
| 0 | Car | 22.01 | 38.28 | 32.93 | 41.41 | 54.43 |
| | Factory | 10.25 | 35.48 | 30.21 | 46.78 | 52.83 |
| | Pink | 22.80 | 45.18 | 37.26 | 51.86 | 57.94 |
| | White | 26.74 | 48.84 | 43.87 | 53.94 | 57.88 |
| 5 | Car | 25.81 | 61.98 | 54.11 | 66.59 | 73.33 |
| | Factory | 02.69 | 42.80 | 39.02 | 56.92 | 65.94 |
| | Pink | 18.98 | 47.72 | 43.83 | 57.47 | 65.97 |
| | White | 23.95 | 45.95 | 44.02 | 56.66 | 63.41 |

to the relatively moderate improvements in average EER, a significant reduction of average MinDCF value of $7.5 \times 10^{-3}$ is noticed. This is comparatively much higher than the previously recorded average MinDCF reduction (see Table 5.11) of $3 \times 10^{-3}$ due to the effect of UP-AVR.

Table 5.13 summarizes the relative EERs of the various SV systems developed in uniform background environments at 0 and 5 dB SNRs. In order to jointly represent the performances of the KL div and GUMI kernels, the mean of their relative EERs has been recorded under the GMM-SVM multicolumn. The significant performance improvements achieved in each stage of development of the i-vector based SV systems, can be more clearly deduced by the large relative EER metrics. A contrasting behavior in performance improvement of the i-vector based SV systems is observed at the two SNR levels. The colored noises (pink and white) which had comparatively lower performance accuracies, are the ones with higher relative EER improvements at 0 dB SNR. However, the environmental noises (car and factory) perform much better at 5 dB SNR. The average relative EERs (across both SNR levels) of the GMM-SVM based SV systems are 19.15% (without UP-AVR) and 45.78% (with UP-AVR), respectively. The corresponding EER values for the i-vector based SV systems are 40.66 and 53.96%, respectively. The DataBoost-UP algorithm in the i-vector framework outperforms the rest of the methods with an average relative EER of 61.47% across all environments.

## 5.5 Summary

This chapter explored the impact of robust speaker models for speaker verification in various noisy environments. Broadly, two types of hybrid modeling techniques (i.e., GMM-SVM and i-vectors SVM) were used to develop SV systems in uniform and varying background environments, respectively. The majority of studies were concentrated in the GMM-SVM based approach. Through extensive experimentation, it was established that robust SV performance could be achieved using GMM supervectors in a discriminative framework, in comparison to the traditional GMM-UBM framework. In particular, emphasis was laid on the significance of using partitioned utterances, for mitigating data imbalance, utterance-duration mismatch and small sample-size problems, respectively for improving performances in SVM based SV framework. In order to enhance SV performances in highly degraded environments, a low-dimensional channel robust representation of GMM supervectors (namely i-vectors), were alternatively used in a SVM framework. A novel boosting algorithm was proposed to address some inherent drawbacks in the standard utterance partitioning scheme and strengthening the SVM classification accuracy in highly degraded background environments.

# References

1. D. Reynolds, T. Quatieri, R. Dunn, Speaker verification using adapted Gaussian mixture models. Digit. Signal Process. **10**(1), 19–41 (2000)
2. W. Campbell, J. Campbell, D. Reynolds, Support vector machines using GMM supervectors for speaker verification. IEEE Signal Process. Lett. **13**(5), 308–311 (2006)
3. W. Campbell, J. Campbell, D. Reynolds, E. Singer, P. Carrasquillo, Support vector machines for speaker and language recognition. Comput. Speech Lang. **20**, 210–229 (2006)
4. C.M. Bishop, *Pattern Recognition and Machine Learning* (Springer, New York, 2006)
5. V. Wan, S.Renals, Speaker verification using sequence discriminant support vector machines. IEEE Trans. Acoust. Speech Audio Process. **13**(2), 203–210 (2005)
6. P. Kenny, G. Boulianne, P. Dumouchel, Eigenvoice modeling with sparse training data. IEEE Trans. Speech Audio Process. **13**(3), 345–354 (2005)
7. N. Dehak, P.J. Kenny, R. Dehak, P. Dumouchel, P. Ouellet, Front-end factor analysis for speaker verification. IEEE Trans. Audio Speech Lang. Process. **19**(4), 788–798 (2011)
8. S. Sarkar, K.S. Rao, Speaker verification in noisy environment using GMM supervectors, in *National Conference on Communications (NCC)*, Delhi (IIT Delhi, Delhi, 2013)
9. C.H. You, K.A. Lee, H. Li, An SVM kernel with GMM-Supervector based on the Bhattacharyya distance for speaker recognition. IEEE Signal Process. Lett. **16**(1), 49–52 (2009)
10. A. Solomonoff, C. Quillen, I. Boardman, Channel compensation for SVM speaker recognition, in *IEEE Workshop on Speaker and Language Recognition (Odyssey '04)*, Toledo, 2004, pp. 57–62
11. R. Akbani, S. Kwek, N. Japkowicz, Applying support vector machines to imbalanced datasets, in *Proceedings of the 15th European Conference on Machine Learning*, Pisa, 2004, vol. 3201, pp. 39–50
12. Y. Tang, Y.Q. Zhang, N.V. Chawla, S. Krasser, SVMs modeling for highly imbalanced classification. IEEE Trans. Syst. Man Cybern. Part B Cybern. **39**, 281–288 (2009)
13. J. Pelecanos, U. Chaudhari, G. Ramaswamy, Compensation of utterance length for speaker verification, in *ODYSSEY04 – The Speaker and Language Recognition Workshop*, Toledo, 2004
14. B. Fauve, N. Evans, J. Mason, Improving the performance of text-independent short duration SVM- and GMM-based speaker verification, in *Workshop on Speaker and Language Recognition (Odyssey)*, Stellenbosch, 2008
15. W. Rao, M.W. Mak, Boosting the performance of i-vector based speaker verification via utterance partitioning. IEEE Trans. Audio Speech Lang. Process. **21**(5), 1012–1022 (2013)
16. L.F. Chen, H.Y.M. Liao, M.T. Ko, J.C. Lin, G.J. Yu, A new LDA-based face recognition system which can solve the small sample size problem. Pattern Recognit. **33**(10), 1713–1726 (2000)

17. J. Ye, Characterization of a family of algorithms for generalized discriminant analysis on undersampled problems. J. Mach. Learn. Res. **6**, 483–502 (2005)
18. N. Sen, H. Patil, S.K.D. Mandal, K.S. Rao, Importance of utterance partitioning in SVM classifier with GMM supervectors for text-independent speaker verification, in *Mining Intelligence and Knowledge Exploration*. LNCS (Springer, Cham, 2013), pp. 780–789
19. N. Chawla, K. Bowyer, L. Hall, W.P. Kegelmeyer, SMOTE: synthetic minority over-sampling technique. J. Artif. Intell. Res. **16**, 341–378 (2002)
20. N. Chawla, A. Lazarevic, L. Hall, K. Bowyer, SMOTEBoost: improving prediction of the minority class in boosting, in *7th European Conference on Principles and Practice of Knowledge Discovery in Databases*, Cavtat-Dubrovnik, 2003
21. P. Kang, S. Cho, EUS SVMs: Ensemble of under-sampled SVMs for data imbalance problems, in *ICONIP (1)*, Hong Kong, 2006, pp. 837–846
22. Z. Lin, Z. Hao, X. Yang, X. Liu, Several SVM ensemble methods integrated with under-sampling for imbalanced data learning, in *Advanced Data Mining and Applications* (Springer, Berlin/Heidelberg, 2009), pp. 536–544
23. K. Veropoulos, C. Campbell, N. Cristianini, Contolling the sensitivity of support vector machines, in *Proceedings of International Joint Conference on Artificial Intelligence*, Stockholm, 1999
24. G. Wu, E. Chang, KBA: kernel boundary alignment considering imbalanced data distribution. IEEE Trans. Knowl. Data Eng. **17**(6), 786–795 (2005)
25. M.W. Mak, W. Rao, Utterance partitioning with acoustic vector resampling for GMM-SVM speaker verification. Speech Commun. **53**, 119–130 (2011)
26. S. Sarkar, K.S. Rao, Significance of utterance partitioning in GMM-SVM based speaker verification in varying background environment, in *16th International Oriental COCOSDA Conference*, Gurgoan, 2013
27. P. Kenny, G. Boulianne, P. Ouellet, P. Dumouchel, Speaker and session variability in GMM-based speaker verification. IEEE Trans. Audio Speech Lang. Process. **15**(4), 1448–1460 (2007)
28. P. Kenny, G. Boulianne, P. Ouellet, P. Dumouchel, Factor analysis simplified, in *Proceedings of the IEEE International Conference on Acoustics Speech and Signal Processing (ICASSP '05)*, Philadelphia, 2005, vol. 1, pp. 637–640
29. N. Dehak, P. Kenny, R. Dehak, O. Glembek, P. Dumouchel, L. Burget, V. Hubeika, F. Castaldo, Support vector machines and joint factor analysis for speaker verification, in *Proceedings of IEEE International Conference on Acoustics, Speech and Signal Processing (ICASSP '09)*, Taipei, 2009, pp. 4237–4240
30. N. Dehak, R. Dehak, P. Kenny, N. Brummer, P. Ouellet, P. Dumouchel, Support vector machines versus fast scoring in the low-dimensional total variability space for speaker verification, in *Proceeding of the 10th Annual Conference of the International Speech Communication Association (INTERSPEECH '09)*, Brighton, 2009
31. A.O. Hatch, S. Kajarekar, A. Stolcke, Within-class covariance normalization for SVM-based speaker recognition, in *Proceedings of the International Conference of Spoken Language Processing (ICSLP '05)*, Jeju, 2005
32. Y. Freund, R. Schapire, Experiments with a new boosting algorithm, in *Proceedings of Thirteenth International Conference on Machine Learning (ICML '96)*, Bari, 1996
33. A. Roy, M.M. Doss, S. Marcel, Boosted binary features for noise-robust speaker verification, in *Proceedings of IEEE International Conference on Acoustics, Speech and Signal Processing (ICASSP '10)*, Dallas, 2010, pp. 4442–4445
34. Y. Sun, S. Todorovic, J. Li, Reducing the overfitting of AdaBoost by controlling its data distribution skewness. Int. J. Pattern Recognit. Artif. Intell. **20**, 1093–1116 (2006)
35. H. Guo, H.L. Viktor, Learning from imbalanced data sets with boosting and data generation: The DataBoost-IM approach. ACM SIGKDD Explor. Newsl. Spec. Issue Learn. Imbalanced Datasets **6**, 30–39 (2004)

# Chapter 6
# Summary and Conclusion

**Abstract** This chapter summarizes the research work presented in this book. It highlights the contributions of the work, and briefly mention the scope for future work.

## 6.1 Summary of the Book

The book addressed the issue of speaker verification (SV) in noisy background environments. It was observed that background noise (in general additive in nature) severely degraded the performance of standard GMM-UBM based SV systems which have been considered as baseline for comparison in the present work. Performance degradation was found to occur in both matched and mismatched conditions, with much inferior performance in the latter situation. A review of conventional strategies for handling background noise revealed two broad categories of methods viz. (i) compensation or adaptation schemes in which speaker model parameters learned in one environment are 'altered' or 'adapted' to reflect the changes in another one and (ii) extraction of 'high-level' features which are inherently robust towards noise distortions. The drawbacks associated with the first category of methods are requirement of clean speaker models, a statistical noise model and high amount of adaptation data while those with the second category is low recognition power.

The aim of the present work was to explore alternative methods for developing robust SV systems (a) without compromising on the discriminative power of 'low-level' features and (b) by acoustic modeling methods that do not rely on the availability of clean speaker models, large amount of adaptation data and a priori knowledge about the test environment. The methods explored in the present work can be broadly categorized as (i) feature transformation methods and (ii) robust speaker modeling methods. The first category in turn comprised two methods i.e., (a) feature mapping in a multiple background framework and (b) stochastic feature compensation methods. The significance of each of these methods were analyzed on the basis of their effectiveness and computational costs. In the feature mapping method, noisy utterances were frame-wise transformed to a noise-independent background model space by simple parametric scaling. However, the performance improvements compared to the baseline (GMM-UBM) systems were nominal in both matched and mismatched conditions. Alternatively, integration of stereo-data based stochastic feature compensation (SFC) methods in the GMM-UBM framework was proposed for SV in mismatched environments. Front end GMMs built from stereo-training data were used for transforming features. During the evaluation phase, noisy test utterances were transformed on the basis of a minimum mean squared error (MMSE) or maximum likelihood (MLE) estimate. Joint probability model based SFC methods resulted in significant performance improvements in comparison to the baseline.

Apart from feature transformation, the present work also explored speaker modeling methods that are relatively immune towards noise distortions. The role of GMM supervectors and total variability modeling (i-vectors) were explored for robust speaker modeling in a discriminative framework using SVMs. In order to justify its robustness towards background noise distortions, the models were constructed using noisy-degraded training utterances and evaluated under matched conditions. An extensive set of experiments were conducted using two kinds of background simulations (i.e., uniform and varying), at various SNRs. The impact of an utterance partitioning (UP-AVR) method for SV in noisy environment, was demonstrated. The moderate performance improvements obtained initially from the default system configuration could be scaled drastically by using partitioned enrollment utterances. In contrast, the SV performance improvements under clean conditions due to UP-AVR, were negligible. This further confirmed the effectiveness of UP-AVR specifically for SV in noisy environments. The significance of the UP-AVR strategy for SV in noisy environment could also be established from the enhanced performances obtained via session compensation techniques like NAP, LDA and WCCN. A few typical drawbacks of the conventional UP-AVR based SV was highlighted. As a remedy to these drawbacks and for combining robust SVM classifiers an adaptive boosting algorithm (DataBoost-UP) was proposed. The proposed boosting method was reasonably effective in enhancing performance accuracies of the SV systems in low SNR conditions.

## 6.2  Major Contribution of the Book

The contribution of the book can be broadly summarized under the following points

- Feature compensation using multiple background models has been proposed for SV in noisy background environments.
- Integration of data-driven stochastic feature compensation methods in the GMM-UBM framework has been proposed for improved robust speaker verification (SV) in noisy background environments under mismatched conditions.

- The robustness of GMM-SVM framework for speaker modeling have been explored for SV in noisy environments under matched conditions.
- The robustness of total variability modeling (i-vectors) in a discriminative framework has been explored for SV in noisy environments.
- Utterance partitioning has been proposed for enhancing SV performances in noisy environments.
- A boosting algorithm has been proposed for combining robust SVM classifiers for improving SV performance.

## 6.3 Scope for Future Work

The following areas can be further explored for robust speaker verification

- Performance of the SV systems in mismatch condition caused due to noisy training environment and clean test environment may be explored.
- In the feature mapping framework discussed in Chap. 3, speech data corrupted with additive noises can be used for constructing multiple background model.
- A priori knowledge of the test environment assumed during the training stages of stochastic feature compensation (SFC) methods can be avoided i.e., SV systems trained using stereo data in one environment may be evaluated with test data corrupted in a different background environment.
- SFC based SV methods may be examined for varying background noisy environments by varying noise types and noise strengths.
- The SV systems developed using supervector-based speaker modeling methods, can be evaluated under mismatched conditions.
- The effect of Nuisance Attribute Projection (NAP) can be also be explored for SV systems developed in uniform background environments.
- The effect of eigenchannel compensation methods [1] can be explored for supervector-based modeling framework for SV in noisy environments.
- The role of high-level features (e.g., prosodic features, idiolect features) can be explored for SV in noisy environments.
- Model adaptation methods can be explored for SV in mismatched environments.
- The proposed SV experiments can be carried out using more recent versions of NIST-SRE databases for a better understanding of their overall utility.
- Proposed SFC based SV systems and Robust speaker models may be examined for other real-life environmental noises e.g., street noise, train noise, restaurant noise, babble noise etc.
- Speech enhancement methods may be examined for improving the SV performance in noisy conditions.
- Speech data corrupted in real-life environments can be used instead of simulated noisy data for experimental purpose.

# Reference

1. L. Burget, P. Matejka, P. Schwarz, O. Glembek, J. Cernocký, Analysis of feature extraction and channel compensation in a GMM speaker recognition system. IEEE Trans. Audio Speech Lang. Process. **15**(7), 1979–1986 (2007)

# Appendix A
# Mathematical Details of Stochastic Feature Compensation Methods

The Stochastic feature compensation methods had been introduced in Chap. 1 and implemented for robust speaker verification in Chap. 4. In this Appendix, we provide detailed derivations of the Minimum Mean Squared Error (MMSE) estimator used for the RATZ, SPLICE and MMCN algorithm used in Chap. 4. We also provide details of the EM algorithm used for RATZ, SPLICE and TRAJMAP.

## A.1 MMSE Estimators for Feature Compensation

Depending on the feature compensation algorithm, the noisy and clean feature spaces (stereo-data) are modeled individually using GMMs during the training phase. Given a sequence of $\mathbf{T}$ noisy MFFC vectors $Y = [y_1, y_2, \ldots y_{\mathbf{T}}]$ and clean MFCC vectors $X = [x_1, x_2, \ldots x_{\mathbf{T}}]$ extracted from the stereo training data, the models are given by

$$p(x) = \sum_{j=1}^{M} w_x(j) \mathcal{N}_x(x_t; \mu_x(j), \Sigma_x(j)) \tag{A.1}$$

$$p(y) = \sum_{j=1}^{M} w_y(j) \mathcal{N}_y(y_t; \mu_y(j), \Sigma_y(j)) \tag{A.2}$$

where $w(j)$, $\mu(j)$ and $\Sigma(j)$ denotes the weight, mean vector and covariance matrix of the $j$th multivariate Gaussian component and $M$ is the total number of components. During evaluation, given a noisy test feature vector $y_t$, the MMSE estimator is used to generate an estimated clean vector $\hat{x}_t$ as follows

$$\hat{x}_t = E[x|y_t] = \int_X x p(x|y_t) \mathrm{d}x \tag{A.3}$$

K.S. Rao and S. Sarkar, *Robust Speaker Recognition in Noisy Environments*, SpringerBriefs in Electrical and Computer Engineering, DOI 10.1007/978-3-319-07130-5, © The Author(s) 2014

In the following subsections, we derive the MMSE estimates for RATZ, SPLICE and MEMLIN (multi-environment version of MMCN) algorithms, respectively.

## A.1.1   RATZ

The RATZ algorithm models the clean feature space $p(x)$. It approximates noisy GMM distribution $p(y)$ using corrective vectors $r_j$ and $R_j$ as follows

$$p(y) = \sum_{j=1}^{M} w_y(j) \mathcal{N}_y(y; \mu_x(j) + r_j, \Sigma_x(j) + R_j) \qquad (A.4)$$

Given a test feature vector $y_t$, the MMSE estimated clean vector is given by

$$
\begin{aligned}
\hat{x}_t = E[x|y_t] = \int_X xp(x|y_t)\mathrm{d}x &= \int_X (y_t - r(x))p(x|y_t)\mathrm{d}x \\
&= \int_X y_t\, p(x|y_t)\mathrm{d}x - \int_X r(x)p(x|y_t)\mathrm{d}x \\
&= y_t \int_X p(x|y_t)\mathrm{d}x - \int_X \sum_{j=1}^{M} r(x)p(x, j|y_t)\mathrm{d}x \\
&= y_t - \sum_{j=1}^{M} p(j|y_t) \int_X r(x)p(x|j, y_t)\mathrm{d}x \\
&= y_t - \sum_{j=1}^{M} p(j|y_t)r_j \int_X p(x|j, y_t)\mathrm{d}x \\
&= y_t - \sum_{j=1}^{M} p(j|y_t)r_j \qquad (A.5)
\end{aligned}
$$

In the sixth step it is assumed that $r(x)$ remains constant over the integral and can be approximated by $r_j$.

## A.1.2   SPLICE

The SPLICE algorithm models the noisy feature space $p(y)$. It approximates the clean conditional distribution $p(x|y_t)$ using corrective vectors $r_j$ and $\Gamma_j$ as follows

$$p(x|y_t) = \sum_{j=1}^{M} w_y(j) \mathcal{N}_y(x; y_t + r_j, \Gamma_j) \tag{A.6}$$

Given a test feature vector $y_t$, the MMSE estimated clean vector is given by

$$\begin{aligned}
\hat{x}_t = E[x|y_t] &= \int_X x p(x|y_t) dx \\
&= \int_X \sum_{j=1}^{M} x p(j|y_t) p(x|j, y_t) dx \\
&= \sum_{j=1}^{M} p(j|y_t) \int_X x p(x|j, y_t) dx \\
&= \sum_{j=1}^{M} p(j|y_t)(y_t + r_j) \\
&= \sum_{j=1}^{M} p(j|y_t) y_t + \sum_{j=1}^{M} p(j|y_t) r_j \\
&= y_t + \sum_{j=1}^{M} p(j|y_t) r_j \tag{A.7}
\end{aligned}$$

### A.1.3 MEMLIN

The MEMLIN algorithm models the noisy feature space $p_e(y)$ for environment $e$, clean feature space $p(x)$ and clean conditional distribution $p(x|y_t, s_y^e, s_x)$ using corrective vectors $r(s_x, s_y^e)$ as follows

$$p_e(y) = \sum_{s_y^e=1}^{M} p(y_t|s_y^e) p(s_y^e)$$

$$p(x) = \sum_{s_x=1}^{M} p(x_t|s_x) p(s_x)$$

$$p(x|y_t) = \sum_{j=1}^{M} w_x(j) \mathcal{N}_x(x; y_t + r_j, \Gamma_j) \tag{A.8}$$

Given a test feature vector $y_t$, the MMSE estimated clean vector is given by

$$\hat{x}_t = E[x|y_t] = \int_x x p(x|y_t) dx$$

$$= \int \sum_e \sum_{s_y^e=1}^{M} \sum_{s_x=1}^{M} x p(x|y_t, s_y^e, s_x, e) dx$$

$$= y_t - \int \sum_e \sum_{s_y^e=1}^{M} \sum_{s_x=1}^{M} r(s_x, s_y^e) p(x|y_t, s_y^e, s_x, e) dx$$

$$= y_t - \sum_e \sum_{s_y^e=1}^{M} \sum_{s_x=1}^{M} r(s_x, s_y^e) p(e|y_t) p(s_y^e|y_t) p(s_x|s_y^e, y_t, e) \quad (A.9)$$

where $r(s_x, s_y^e)$ denotes the additive bias term for noisy GMM component $s_y^e$ and clean GMM component $s_x$, respectively. The noisy training environment is indexed by $e$ while $p(s_x|s_y^e, y_t, e)$ denotes the cross probability model.

## A.2   Expectation Maximization Algorithms for Feature Compensation

The Expectation Maximization (EM) algorithm determines the unknown parameters of a statistical model by iteratively maximizing the likelihood of 'complete data'. The complete data consists of the observed variables (training data) and latent variables (GMM component). The EM algorithm consists of two stages i.e., Expectation (E) and Maximization (M). In the E step, the expected value of a log-likelihood function with respect to the posterior probability of a latent GMM component, is calculated as follows

$$Q(\phi, \phi') = \sum_z p(z|X, \phi') \log (p(z, X|\phi)) \quad (A.10)$$

where $z$ is the unobserved/latent variable, $\phi$ is the estimated parameter set from a previous iteration and $\phi'$ is the set of parameters to be estimated in the current iteration and $Q$ is commonly termed as auxillary function.

In the M step, a new set of parameters are obtained by maximizing the auxillary function as follows

$$\phi = \arg max_\phi Q(\phi, \phi') \quad (A.11)$$

In the following subsections we describe the EM algorithm used for determining the parameters of RATZ, SPLICE and TRAJMAP respectively,

## A.2.1 RATZ

The complete log-likelihood function for RATZ (Sect. A.1.1) is given as follows

$$L(Y) = \log \prod_{t=1}^{T} p(y_t) = \sum_{t=1}^{T} \log \sum_{j=1}^{M} w_y(j) \mathcal{N}_y(y_t; \mu_y(j), \Sigma_y(j))$$

$$= \sum_{t=1}^{T} \log \sum_{j=1}^{M} w_y(j) \mathcal{N}_y(y_t; \mu_x(j) + r_j, \Sigma_x(j) + R_j)$$

$$(A.12)$$

We define the set of unknown parameters to be estimated as the collection of the additive bias terms i.e., $\phi = \{r_1, r_2 \ldots r_M, R_1, R_2 \ldots R_M\}$ and the latent variable for $y_t$ as $s_t(j)$

**E Step:**

$$Q(\phi, \phi') = \sum_{t=1}^{T} \sum_{j=1}^{M} p(s_t(j)|y_t, \phi) \log(p(y_t, s_t(j)|\phi')) \qquad (A.13)$$

$$Q = \sum_{t=1}^{T} \sum_{j=1}^{M} p(s_t(j)|y_t, \phi) \left\{ -\frac{D}{2} \log |R_j + \Sigma_x(j)| - \right.$$

$$\left. \frac{1}{2} (y_t - \mu_x(j) - \hat{r}_j)^T (\Sigma_x(j) + R_j)^{-1} (y_t - \mu_x(j) - \hat{r}_j) \right\} + K$$

where $Q = Q(\phi, \phi')$, '$D$' is the dimension of the feature space and '$K$' is a constant term independent of the bias parameters.

**M Step:**

$$\frac{\partial Q}{\partial \hat{r}_j} = \sum_{t=1}^{T} p(s_t(j)|y_t, \phi)(\Sigma_x(j) + R_j)^{-1} (y_t - \mu_x(j) - \hat{r}_j) = 0$$

$$=> \sum_{t=1}^{T} p(s_t(j)|y_t, \phi)(y_t - \mu_x(j) - \hat{r}_j) = 0$$

$$=> \hat{r}_j = \frac{\sum_{t=1}^{T} p(s_t(j)|y_t, \phi)(y_t - \mu_x(j))}{\sum_{t=1}^{T} p(s_t(j)|y_t, \phi)} \qquad (A.14)$$

The expression for $\Sigma_x(j)$ can be obtained likewise.

### A.2.2    SPLICE

The complete log-likelihood function for SPLICE (Sect. A.1.2) is given as follows

$$L(X) = \sum_{t=1}^{T} \log(p(x_t))$$

$$= \sum_{t=1}^{T} \log \sum_{j=1}^{M} w_x(j) \mathcal{N}_x(x_t; \mu_x(j), \Sigma_x(j))$$

$$= \sum_{t=1}^{T} \log \sum_{j=1}^{M} w_y(j) \mathcal{N}_y(x_t; y_t + r_j, \Gamma_j) \qquad (A.15)$$

We define the set of unknown parameters to be estimated as the collection of the additive bias terms i.e., $\phi = \{r_1, r_2 \ldots r_M, \Gamma_1, \Gamma_2 \ldots \Gamma_M\}$ and the latent variable for $x_t$ as $s_t(j)$

**E Step:**

$$Q(\phi, \phi') = \sum_{t=1}^{T} \sum_{j=1}^{M} p(s_t(j)|y_t, \phi) \log(p(x_t, s_t(j)|\phi')) \qquad (A.16)$$

$$Q = \sum_{t=1}^{T} \sum_{j=1}^{M} p(s_t(j)|y_t, \phi) \left\{ -\frac{D}{2} \log |\Gamma_j| - \frac{1}{2}(x_t - y_t - \hat{r}_j)^T (\Gamma_j)^{-1}(x_t - y_t - \hat{r}_j) \right\} + K$$

where $Q=Q(\phi, \phi')$, '$D$' is the dimension of the feature space and '$K$' is a constant term independent of the bias parameters.

**M Step:**

$$\frac{\partial Q}{\partial \hat{r}_j} = \sum_{t=1}^{T} p(s_t(j)|y_t, \phi)(\Gamma_j)^{-1}(x_t - y_t - \hat{r}_j) = 0$$

$$=> \sum_{t=1}^{T} p(s_t(j)|y_t, \phi)(x_t - y_t - \hat{r}_j) = 0$$

$$=> \hat{r}_j = \frac{\sum_{t=1}^{T} p(s_t(j)|y_t, \phi)(x_t - y_t)}{\sum_{t=1}^{T} p(s_t(j)|y_t, \phi)} \tag{A.17}$$

For notational convenience $s_t(j)$ have been replaced by '$j$' and $\phi$ has been omitted

$$=> \hat{r}_j = \frac{\sum_{t=1}^{T} p(j|y_t)(x_t - y_t)}{\sum_{t=1}^{T} p(j|y_t)} \tag{A.18}$$

The expression for $\Gamma_j$ can be obtained likewise.

## A.2.3   TRAJMAP

The complete likelihood function for the TRAJMAP algorithm is given by

$$p(\mathbf{X}|\mathbf{Y}, \lambda^{(Z)}) = \sum_{\mathbf{j}} p(\mathbf{j}|\mathbf{Y}, \lambda^{(Z)}) p(\mathbf{X}|\mathbf{Y}, \mathbf{j}, \lambda^{(Z)})$$

$$= \prod_{t=1}^{T} \sum_{j=1}^{M} p(j|Y_t, \lambda^{(Z)}) p(X_t|Y_t, j, \lambda^{(Z)}) \tag{A.19}$$

where $\mathbf{j} = \{j_1, j_2 \ldots j_T\}$ is a mixture component sequence. The conditional pdf at each frame is modeled as a GMM. At frame $t$, the $j$th mixture component weight $p(j|Y_t, \lambda^{(Z)})$ and the $j$th conditional probability distribution $p(X_t|Y_t, j, \lambda^{(Z)})$ are given by the following expressions

$$p(j|Y_t, \lambda^{(Z)}) = \frac{w_j^Y \mathcal{N}(Y_t; \mu_j^Y, \Sigma_j^{YY})}{\sum_{j=1}^{M} w_j^Y \mathcal{N}(Y_t; \mu_j^Y, \Sigma_j^{YY})} \tag{A.20}$$

$$p(X_t|Y_t, j, \lambda^{(Z)}) = \mathcal{N}(X_t; E_{j,t}^X, D_j^X) \tag{A.21}$$

where

$$E_{j,t}^X = \mu_j^X + \Sigma_j^{XY}(\Sigma_j^{YY})^{-1}(Y_t - \mu_j^Y) \tag{A.22}$$

$$D_j^X = \Sigma_j^{XX} - \Sigma_j^{XY}(\Sigma_j^{YY})^{-1}\Sigma_j^{YX} \tag{A.23}$$

The task is to estimate a static feature vector sequence $\hat{x} = [\hat{x}_1^T, \hat{x}_2^T, \ldots, \hat{x}_T^T]$ by maximizing the likelihood function given by Eq. (A.19) ($\hat{x} = \arg\max\ p(\mathbf{X}|\mathbf{Y}, \lambda^{(Z)})$). This in turn is achieved by iteratively maximizing the auxillary function $Q(\mathbf{X}, \hat{\mathbf{X}})$ using an EM algorithm as follows

**E Step:**

$$Q(\mathbf{X}, \hat{\mathbf{X}}) = \sum_{\mathbf{j}} p(\mathbf{j}|\mathbf{Y}, \mathbf{X}, \lambda^{(Z)}) \log(p(\hat{\mathbf{X}}, \mathbf{j}|\mathbf{Y}, \lambda^{(Z)}))$$

$$= \sum_{t=1}^{T} \sum_{j=1}^{M} p(j|Y_t, X_t, \lambda^{(Z)}) \log(p(\hat{X}_t, j|Y_t, \lambda^{(Z)}))$$

$$= \sum_{t=1}^{T} \sum_{j=1}^{M} \lambda_{j,t} (-\frac{1}{2}\hat{X}_t^T (D_j^X)^{-1} \hat{X}_t + \hat{X}_t^T (D_j^X)^{-1} E_{j,t}^X) + K$$

$$= \sum_{t=1}^{T} (-\frac{1}{2} \sum_{j=1}^{M} \lambda_{j,t} \hat{X}_t^T (D_j^X)^{-1} \hat{X}_t + \sum_{j=1}^{M} \lambda_{j,t} \hat{X}_t^T (D_j^X)^{-1} E_{j,t}^X) + K$$

$$= \sum_{t=1}^{T} (-\frac{1}{2} \hat{X}_t^T \overline{(D_t^X)^{-1}} \hat{X}_t + \hat{X}_t^T \overline{(D_t^X)^{-1} E_t^X}) + K$$

$$= -\frac{1}{2} \hat{X}^T \overline{(\mathbf{D^X})^{-1}} \hat{X} + \hat{X}^T \overline{(\mathbf{D^X})^{-1} \mathbf{E^X}} + K$$

$$= -\frac{1}{2} \hat{x}^T \mathbf{W}^T \overline{(\mathbf{D^X})^{-1}} \mathbf{W} \hat{x} + \hat{x}^T \mathbf{W}^T \overline{(\mathbf{D^X})^{-1} \mathbf{E^X}} + K \qquad \text{(A.24)}$$

where $K$ in the third step of the above derivation is a constant term independent of $\hat{X}$, $\hat{X} = \mathbf{W}\hat{x}$ and

$$\overline{(\mathbf{D^X})^{-1}} = diag[\overline{(D_1^X)^{-1}}, \overline{(D_2^X)^{-1}}, \ldots, \overline{(D_t^X)^{-1}}, \ldots, \overline{(D_T^X)^{-1}}] \qquad \text{(A.25)}$$

$$\overline{(\mathbf{D^X})^{-1} \mathbf{E^X}} = [\overline{(D_1^X)^{-1} E_1^X}^T, \overline{(D_2^X)^{-1} E_2^X}^T, \ldots, \overline{(D_t^X)^{-1} E_t^X}^T, \ldots, \overline{(D_T^X)^{-1} E_T^X}^T]^T \qquad \text{(A.26)}$$

$$\overline{(D_t^X)^{-1}} = \sum_{j=1}^{M} \lambda_{j,t} (D_j^X)^{-1} \qquad \text{(A.27)}$$

$$\overline{(D_t^X)^{-1} E_t^X} = \sum_{j=1}^{M} \lambda_{j,t} (D_j^X)^{-1} E_{j,t}^X \qquad \text{(A.28)}$$

$$\lambda_{j,t} = p(j|Y_t, X_t, \lambda^{(Z)}) \qquad \text{(A.29)}$$

## M Step:

The partial derivative of the auxillary function with respect to $\hat{x}$ gives the following equation

$$\frac{\partial Q(\mathbf{X}, \hat{\mathbf{X}})}{\partial \hat{x}} = -\mathbf{W}^T \overline{(\mathbf{D}^\mathbf{X})^{-1}} \mathbf{W} \hat{x} + \mathbf{W}^T \overline{(\mathbf{D}^\mathbf{X})^{-1}} \mathbf{E}^\mathbf{X} \tag{A.30}$$

The sequence of vector $\hat{x}$ obtained by setting the above partial derivative to 0 is given by

$$\hat{x} = (\mathbf{W}^T \overline{(\mathbf{D}^\mathbf{X})^{-1}} \mathbf{W})^{-1} \mathbf{W}^T \overline{(\mathbf{D}^\mathbf{X})^{-1}} \mathbf{E}^\mathbf{X} \tag{A.31}$$

# Appendix B
# Gaussian Mixture Model

In speaker recognition, the acoustic events are usually modeled by Gaussian probability density functions (PDFs), described by the mean vector and the covariance matrix. However unimodel PDF with only one mean and covariance are unsuitable to model all variations of a single event in speech signals. Therefore, a mixture of single densities i.e., a Gaussian Mixture Model (GMM) is used to model the complex structure of the density probability. For a $D$-dimensional feature vector denoted as $x_t$, the mixture density for speaker $\Omega$ is defined as weighted sum of $M$ component Gaussian densities as given by the following [1]

$$P(x_t|\Omega) = \sum_{i=1}^{M} w_i P_i(x_t) \tag{B.1}$$

where $w_i$ are the weights and $P_i(x_t)$ are the component densities. Each component density is a $D$-variate Gaussian function of the form

$$P_i(x_t) = \frac{1}{(2\pi)^{D/2}|\Sigma_i|^{\frac{1}{2}}} e^{-\frac{1}{2}[(x_t-\mu_i)'\Sigma_i^{-1}(x_t-\mu_i)]} \tag{B.2}$$

where $\mu_i$ is a mean vector and $\Sigma_i$ covariance matrix for $i$th component. The mixture weights have to satisfy the constraint [1]

$$\sum_{i=1}^{M} w_i = 1. \tag{B.3}$$

The complete Gaussian mixture density is parameterized by the mean vector, the covariance matrix and the mixture weight from all component densities. These parameters are collectively represented by

$$\Omega = \{w_i, \mu_i, \Sigma_i\}; \qquad i = 1, 2, \dots M. \tag{B.4}$$

K.S. Rao and S. Sarkar, *Robust Speaker Recognition in Noisy Environments*,
SpringerBriefs in Electrical and Computer Engineering,
DOI 10.1007/978-3-319-07130-5, © The Author(s) 2014

## B.1   Training the GMMs

To determine the model parameters of GMM of the speaker, the GMM has to be trained. In the training process, the maximum likelihood (ML) procedure is adopted to estimate model parameters. For a sequence of training vectors $X = \{x_1, x_2, \ldots, x_T\}$, the GMM likelihood (assuming independent observations) can be written as [1]

$$P(X|\Omega) = \prod_{t=1}^{T} P(x_t|\Omega). \tag{B.5}$$

Usually this is done by taking the logarithm and is commonly named as log-likelihood function. From Eqs. (B.1) and (B.5), the log-likelihood function can be written as

$$\log [P(X|\Omega)] = \sum_{t=1}^{T} \log \left[ \sum_{i=1}^{M} w_i P_i(x_t) \right]. \tag{B.6}$$

Often, the average log-likelihood is used value is used by dividing $\log [P(X|\Omega)]$ by $T$. This is done to normalize out duration effects from the log-likelihood value. Also, since the incorrect assumption of independence is underestimating the actual likelihood value with dependencies, scaling by $T$ can be considered a rough compensation factor [2]. The parameters of a GMM model can be estimated using maximum likelihood (ML) estimation. The main objective of the ML estimation is to derive the optimum model parameters that can maximize the likelihood of GMM. The likelihood value is, however, a highly nonlinear function in the model parameters and direct maximization is not possible. Instead, maximization is done through iterative procedures. Of the many techniques developed to maximize the likelihood value, the most popular is the iterative expectation maximization (EM) algorithm [3].

### B.1.1   Expectation Maximization (EM) Algorithm

The EM algorithm begins with an initial model $\Omega$ and tends to estimate a new model such that the likelihood of the model increasing with each iteration. This new model is considered to be an initial model in the next iteration and the entire process is repeated until a certain convergence threshold is obtained or a certain predetermined number of iterations have been made. A summary of the various steps followed in the EM algorithm are described below.

1. **Initialization:** In this step an initial estimate of the parameters is obtained. The performance of the EM algorithm depends on this initialization. Generally, LBG [4] or K-means algorithm [5] is used to initialize the GMM parameters.
2. **Likelihood Computation:** In each iteration the posterior probabilities for the $i$th mixture is computed as [1]:

$$\Pr(i|x_t) = \frac{w_i\, P_i(x_t)}{\displaystyle\sum_{j=1}^{M} w_j\, P_j(x_t)}.$$  (B.7)

3. **Parameter Update:** Having the posterior probabilities, the model parameters are updated according to the following expressions [1].
   Mixture weight update:

$$\overline{w_i} = \frac{\displaystyle\sum_{i=1}^{T} \Pr(i|x_t)}{T}.$$  (B.8)

Mean vector update:

$$\overline{\mu_i} = \frac{\displaystyle\sum_{i=1}^{T} \Pr(i|x_t)x_t}{\displaystyle\sum_{i=1}^{T} \Pr(i|x_t)}.$$  (B.9)

Covariance matrix update:

$$\overline{\sigma}_i^2 = \frac{\displaystyle\sum_{i=1}^{T} \Pr(i|x_t)\,|x_t - \overline{\mu_i}|^2}{\displaystyle\sum_{i=1}^{T} \Pr(i|x_t)}.$$  (B.10)

In the estimation of the model parameters, it is possible to choose, either full covariance matrices or diagonal covariance matrices. It is more common to use diagonal covariance matrices for GMM, since linear combination of diagonal covariance Gaussians has the same model capability with full matrices. Another reason is that speech utterances are usually parameterized with cepstral features. Cepstral features are more compactable, discriminative, and most important, they are nearly uncorrelated, which allows diagonal covariance to be used by the GMMs [1]. The iterative process is normally carried out 10 times, at which point the model is assumed to converge to a local maximum [1].

## B.2   Testing

In identification phase, mixture densities are calculated for every feature vector for all speakers and speaker with maximum likelihood is selected as identified speaker. For example, if $S$ speaker models $\{\Omega_1, \Omega_2, \ldots, \Omega_S\}$ are available after the training, speaker identification can be done based on a new speech data set. First, the sequence of feature vectors $X = \{x_1, x_2, \ldots, x_T\}$ is calculated. Then the speaker model $\hat{s}$ is determined which maximizes the a posteriori probability $P\,(\Omega_S|X)$. That is, according to the Bayes rule [1]

$$\hat{s} = \max_{1 \leq s \leq S} P\,(\Omega_S|X) = \max_{1 \leq s \leq S} \frac{P\,(X|\Omega_S)}{P(X)} P(\Omega_S). \tag{B.11}$$

Assuming equal probability of all speakers and the statistical independence of the observations, the decision rule for the most probable speaker can be redefined as

$$\hat{s} = \max_{1 \leq s \leq S} \sum_{t=1}^{T} \log P(x_t|\Omega_s) \tag{B.12}$$

with $T$ the number of feature vectors of the speech data set under test and $P(x_t|\Omega_s)$ given by Eq. (B.1).

Decision in verification is obtained by comparing the score computed using the model for the claimed speaker $\Omega_S$ given by $P\,(\Omega_S|X)$ to a predefined threshold $\theta$. The claim is accepted if $P\,(\Omega_S|X) > \theta$, and rejected otherwise [6].

## References

1. D.A. Reynolds, R.C. Rose, Robust text-independent speaker identification using Gaussian mixture speaker models. IEEE Trans. Acoust. Speech Signal Process. **3**(1), 72–83 (1995)
2. F. Bimbot, J.F. Bonastre, C. Fredouille, G. Gravier, I. Magrin-Chagnolleau, S. Meignier, T. Merlin, J. Ortega-Garcia, D. Petrovska-Delacrétaz, D.A. Reynolds, A tutorial on text-independent speaker verification. EURASIP J. Adv. Signal Process. (Spec. Issue Biom. Signal Process.) **4**(4), 430–451 (2004)
3. A. Dempster, N. Laird, D. Rubin, Maximum likelihood from incomplete data via the EM algorithm. J. R. Stat. Soc. **39**(1), 1–38 (1977)
4. Y. Linde, A. Buzo, R. Gray, An algorithm for vector quantizer design. IEEE Trans. Commun. **28**, 84–95 (1980)
5. C.M. Bishop, *Pattern Recognition and Machine Learning* (Springer, New York, 2006)
6. D. Reynolds, T. Quatieri, R. Dunn, Speaker verification using adapted Gaussian mixture models. Digit. Signal Process. **10**(1), 19–41 (2000)

# Appendix C
# MFCC Features

The MFCC feature extraction technique basically includes windowing the signal, applying the DFT, taking the log of the magnitude and then warping the frequencies on a Mel scale, followed by applying the inverse DCT. The detailed description of various steps involved in the MFCC feature extraction is explained below.

1. **Pre-emphasis:** Pre-emphasis refers to filtering that emphasizes the higher frequencies. Its purpose is to balance the spectrum of voiced sounds that have a steep roll-off in the high frequency region. For voiced sounds, the glottal source has an approximately $-12$ dB/octave slope [1]. However, when the acoustic energy radiates from the lips, this causes a roughly $+6$ dB/octave boost to the spectrum. As a result, a speech signal when recorded with a microphone from a distance has approximately a $-6$ dB/octave slope downward compared to the true spectrum of the vocal tract. Therefore, pre-emphasis removes some of the glottal effects from the vocal tract parameters. The most commonly used pre-emphasis filter is given by the following transfer function

$$H(z) = 1 - bz^{-1} \tag{C.1}$$

   where the value of $b$ controls the slope of the filter and is usually between 0.4 and 1.0 [1].

2. **Frame blocking and windowing:** The speech signal is a slowly time-varying or quasi-stationary signal. For stable acoustic characteristics, speech needs to be examined over a sufficiently short period of time. Therefore, speech analysis must always be carried out on short segments across which the speech signal is assumed to be stationary. Short-term spectral measurements are typically carried out over 20 ms windows, and advanced every 10 ms [1]. Advancing the time window every 10 ms enables the temporal characteristics of individual speech sounds to be tracked and the 20 ms analysis window is usually sufficient to provide good spectral resolution of these sounds, and at the same time short enough to resolve significant temporal characteristics. The purpose of the overlapping analysis is that each speech sound of the input sequence would be

K.S. Rao and S. Sarkar, *Robust Speaker Recognition in Noisy Environments*,
SpringerBriefs in Electrical and Computer Engineering,
DOI 10.1007/978-3-319-07130-5, © The Author(s) 2014

approximately centered at some frame. On each frame a window is applied to taper the signal towards the frame boundaries. Generally, Hanning or Hamming windows are used [1]. This is done to enhance the harmonics, smooth the edges and to reduce the edge effect while taking the DFT on the signal.

3. **DFT spectrum:** Each windowed frame is converted into magnitude spectrum by applying DFT.

$$X(k) = \sum_{n=0}^{N-1} x(n)e^{\frac{-j2\pi nk}{N}}; \qquad 0 \le k \le N-1 \qquad (C.2)$$

where $N$ is the number of points used to compute the DFT.

4. **Mel-spectrum:** Mel-Spectrum is computed by passing the Fourier transformed signal through a set of band-pass filters known as mel-filter bank. A mel is a unit of measure based on the human ears perceived frequency. It does not correspond linearly to the physical frequency of the tone, as the human auditory system apparently does not perceive pitch linearly. The mel scale is approximately a linear frequency spacing below 1 kHz, and a logarithmic spacing above 1 kHz [1]. The approximation of mel from physical frequency can be expressed as

$$f_{mel} = 2{,}595 \log_{10}\left(1 + \frac{f}{700}\right) \qquad (C.3)$$

where $f$ denotes the physical frequency in Hz, and $f_{mel}$ denotes the perceived frequency [1].

Filter banks can be implemented in both time domain and frequency domain. For MFCC computation, filter banks are generally implemented in frequency domain. The center frequencies of the filters are normally evenly spaced on the frequency axis. However, in order to mimic the human ears perception, the warped axis according to the non-linear function given in Eq. (C.3), is implemented. The most commonly used filter shaper is triangular, and in some cases the Hanning filter can be found [1]. The triangular filter banks with mel-frequency warping is given in Fig. C.1.

The mel spectrum of the magnitude spectrum $X(k)$ is computed by multiplying the magnitude spectrum by each of the of the triangular mel weighting filters.

$$s(m) = \sum_{k=0}^{N-1} \left[|X(k)|^2 \, H_m(k)\right]; \qquad 0 \le m \le M-1 \qquad (C.4)$$

where $M$ is total number of triangular mel weighting filters [2]. $H_m(k)$ is the weight given to the $k$th energy spectrum bin contributing to the $m$th output band and is expressed as:

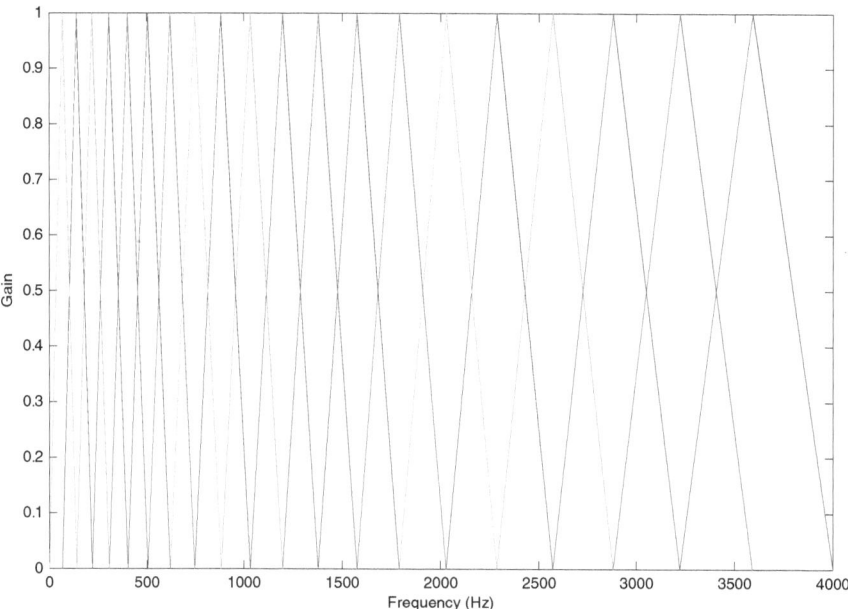

**Fig. C.1** Mel filterbank

$$H_m(k) = \begin{cases} 0, & k < f(m-1) \\ \frac{2(k-f(m-1))}{f(m)-f(m-1)}, & f(m-1) \le k \le f(m) \\ \frac{2(f(m+1)-k)}{f(m+1)-f(m)}, & f(m) < k \le f(m+1) \\ 0, & k > f(m+1) \end{cases} \tag{C.5}$$

with $m$ ranging from 0 to $M-1$.

5. **Discrete Cosine Transform (DCT):** Since the vocal tract is smooth, the energy levels in adjacent bands tend to be correlated. The DCT is applied to the transformed mel frequency coefficients produces a set of cepstral coefficients. Prior to computing DCT the mel spectrum is usually represented on a log scale. This results in a signal in the cepstral domain with a que-frequency peak corresponding to the pitch of the signal and a number of formants representing low quefrequency peaks. Since most of the signal information is represented by the first few MFCC coefficients, the system can be made robust by extracting only those coefficients ignoring or truncating higher order DCT components [1]. Finally, MFCC is calculated as [2]

$$c(n) = \sum_{m=0}^{M-1} \log_{10}(s(m)) \cos\left(\frac{\pi n(m-0.5)}{M}\right); \qquad n = 0, 1, 2, \ldots, C-1 \tag{C.6}$$

where $c(n)$ are the cepstral coefficients and $C$ is the number of MFCCs. Traditional MFCC systems use only 8–13 cepstral coefficients. The zeroth coefficient is often excluded since it represents the average log-energy of the input signal, which only carries little speaker-specific information.

6. **Dynamic MFCC features:** The cepstral coefficients are usually referred to as static features, since they only contain information from a given frame. The extra information about the temporal dynamics of the signal is obtained by computing first and second derivatives of cepstral coefficients [3]. The first order derivative is called delta coefficients, and the second order derivative is called delta-delta coefficients. Delta coefficients tell about the speech rate, and delta-delta coefficients provide information similar to acceleration of speech. The commonly used definition for computing dynamic parameter is

$$\Delta c_m(n) = \frac{\sum\limits_{i=-T}^{T} k_i c_m(n+i)}{\sum\limits_{i=-T}^{T} |i|} \tag{C.7}$$

where $c_m(n)$ denotes the $m$th feature for the $n$th time frame, $k_i$ is the $i$th weight and $T$ is the number of successive frames used for computation. Generally $T$ is taken as 2. The delta-delta coefficients are computed by taking the first order derivative of the delta coefficients.

# References

1. L. Rabiner, B.H. Juang, *Fundamentals of Speech Recognition*, 1st edn. (Prentice-Hall, Englewood Cliffs, 1993)
2. S. Davis, P. Mermelstein, Comparison of parametric representations for monosyllabic word recognition in continuously spoken sentences. IEEE Trans. Acoust. Speech Signal Process. **28**(4), 357–366 (1980)
3. S. Furui, Cepstral analysis technique for automatic speaker verification. IEEE Trans. Acoust. Speech Signal Process. **29**(2), 254–272 (1981)

# Appendix D
# Total Variability Training

Total variability modeling was introduced in Chap. 5. In this method large dimensional supervectors are projected in a low-dimensional subspace (known as 'total variability' space) based on probabilistic principal component analysis, as follows

$$M = m + Tw \tag{D.1}$$

where $M$ is a GMM mean supervector, $m$ is a speaker/channel independent supervector (i.e., the UBM mean supervector), $T$ is low-rank rectangular matrix whose columns consists of eigenvectors of the total variability covariance matrix with largest eigenvalues. $w$ is a random vector having standard Normal distribution, called i-vector (i.e., $w \backsim \mathcal{N}(0, I)$). $M$ is assumed to be Normal distributed with mean $m$ and covariance $TT^T$ (i.e., $M \backsim \mathcal{N}(m, TT^T)$).

This Appendix discusses the training procedure of the total variability ($T$) matrix. As mentioned earlier in Chap. 5, the $T$ matrix is trained offline using a development dataset. The training algorithm implicitly assumes each utterance in the dataset to be produced by a different speaker irrespective of the actual number of speakers. Let the development data consist of **S** training utterances (each from one speaker). The steps of the training procedure are given as follows

1. **Baum-Welch statistics estimation**: Given an utterance ($s$) of **T** frames consisting a sequence of $D$-dimensional acoustic vectors $\{y_1, y_2, \ldots y_{\mathbf{T}}\}$, the Baum-Welch statistics are calculated as follows

$$N_i(s) = \sum_{t=1}^{\mathbf{T}} p(i|y_t, \lambda)$$

$$F_i(s) = \sum_{t=1}^{\mathbf{T}} p(i|y_t, \lambda)(y_t - m_i)$$

K.S. Rao and S. Sarkar, *Robust Speaker Recognition in Noisy Environments*,
SpringerBriefs in Electrical and Computer Engineering,
DOI 10.1007/978-3-319-07130-5, © The Author(s) 2014

$$S_i(s) = diag\left(\sum_{t=1}^{T} p(i|y_t, \lambda)(y_t - m_i)(y_t - m_i)^T\right) \qquad \text{(D.2)}$$

where $p(i|y_t, \lambda)$ is the posterior probability of the $i$th Gaussian component of a UBM $\lambda$ having total $M$ components, which generates vector $y_t$. The mean of the same component is given by $m_i$. $N_i(s)$, $F_i(s)$ and $S_i(s)$ are known as the zeroth order, mean-shifted first order and mean-shifted second order sufficient statistics, respectively. The '$diag$' operation keeps only the diagonal entries and zeros out the other entries.

2. **Expansion of statistics into matrices**: The Baum-Welch statistics estimated in Step 1 are expanded into matrices as follows

$$N(s) = \begin{bmatrix} N_1(s)I & & \\ & \ddots & \\ & & N_M(s)I \end{bmatrix}$$

$$S(s) = \begin{bmatrix} S_1(s) & & \\ & \ddots & \\ & & S_M(s) \end{bmatrix}$$

$$F(s) = \begin{bmatrix} F_1(s) \\ \vdots \\ F_M(s) \end{bmatrix}$$

where $N(s)$, $S(s)$ are block diagonal matrices of size $MD \times MD$, $F(s)$ is a $MD \times 1$ supervector and $I$ is a $D \times D$ identity matrix.

3. **Accumulation of additional statistics across speakers**: The $T$ matrix of appropriate size is initialized randomly. The following statistics are thereafter estimated as follows

$$N_i = \sum_{s=1}^{S} N_i(s)$$

$$A_i = \sum_{s=1}^{S} N_i(s)l^{-1}(s)$$

$$\mathscr{C} = \sum_{s=1}^{S} F(s)\left(l^{-1}(s)T^T \Sigma^{-1} F(s)\right)^T$$

$$N = \sum_{s=1}^{S} N(s) \qquad \text{(D.3)}$$

where $\Sigma$ is a block diagonal matrix whose diagonal blocks consist of the UBM covariance matrices $\Sigma_i$ ($i = 1, 2, \ldots M$) and $l^{-1}(s) = (I + T^T \Sigma^{-1} N(s)T)^{-1}$.

4. **Re-estimation of $T$ matrix**: Following Step 3, the $T$ matrix is re-estimated as follows

$$T = \begin{bmatrix} T_1 \\ \vdots \\ T_M \end{bmatrix} = \begin{bmatrix} A_1^{-1} \mathscr{C}_1 \\ \vdots \\ A_M^{-1} \mathscr{C}_M \end{bmatrix} \tag{D.4}$$

where $\mathscr{C} = \begin{bmatrix} \mathscr{C}_1 \\ \vdots \\ \mathscr{C}_M \end{bmatrix}$.

5. **Optional updation of UBM covariance matrix**: Depending on the amount of available training data, the UBM covariance matrix $(\Sigma)$ used in Step 3 can be updated optionally as follows

$$\Sigma = N^{-1} \left( \left( \sum_{s=1}^{\mathbf{S}} S(s) \right) - diag(\mathscr{C}T^T) \right) \tag{D.5}$$

It can be noted that the second order statistics $S(s)$ are only used in this step. In scenarios where the covariance updation is avoided, estimation of $S(s)$ as in Steps 1 and 2, can be skipped.

6. **Steps 3–4 (or 3–5) are iterated 20–25 times approximately. The re-estimated value of $T$ in Step 4 of each iteration is substituted in the required equations in Step 3.**